AF595050

Magnetic and Spin Devices

Magnetic and Spin Devices

Editors

Viktor Sverdlov
Nuttachai Jutong

MDPI • Basel • Beijing • Wuhan • Barcelona • Belgrade • Manchester • Tokyo • Cluj • Tianjin

Editors

Viktor Sverdlov
Christian Doppler Laboratory for Nonvolatile Magnetoresistive Memory and Logic,Institute for Microelectronics
TU Wien
Vienna
Austria

Nuttachai Jutong
Research Center for Quantum Technology
Chiang Mai University
Chiang Mai
Thailand

Editorial Office
MDPI
St. Alban-Anlage 66
4052 Basel, Switzerland

This is a reprint of articles from the Special Issue published online in the open access journal *Micromachines* (ISSN 2072-666X) (available at: www.mdpi.com/journal/micromachines/special_issues/magnetic_devices).

For citation purposes, cite each article independently as indicated on the article page online and as indicated below:

LastName, A.A.; LastName, B.B.; LastName, C.C. Article Title. *Journal Name* **Year**, *Volume Number*, Page Range.

ISBN 978-3-0365-3842-6 (Hbk)
ISBN 978-3-0365-3841-9 (PDF)

Contents

About the Editors

Viktor Sverdlov Viktor Sverdlov received his Master of Science and Ph.D. degrees in physics from the State University of St. Petersburg, Russia, in 1985 and 1989, respectively. From 1989 to 1999 he worked as a staff research scientist at the V.A.Fock Institute of Physics, St.Petersburg State University. During this time, he visited ICTP (Italy, 1993), the University of Geneva (Switzerland, 1993-1994), the University of Oulu (Finland,1995), the Helsinki University of Technology (Finland, 1996, 1998), the Free University of Berlin (Germany, 1997), and NORDITA (Denmark, 1998). In 1999, he became a staff research scientist at the State University of New York at Stony Brook. He joined the Institute for Microelectronics, Technische Universität Wien, in 2004 and he is currently in a tenure-track position. His scientific interests include device simulations, computational physics, solid-state physics, and nanoelectronics.

Nuttachai Jutong Nuttachai Jutong received his Master of Science and Ph.D. degrees in electrical engineering from Khon Kaen University, Thailand, in 2005 and 2010, respectively. From 2011 to the end of 2012 and in 2014, he worked as a postdoctoral researcher at the Theoretical Physics II, Institute of Physics, University of Augsburg, Germany, and I.PhysikalischesInstitut, Justus-Liebig-University Giessen, Giessen, Germany, respectively. From 2005 to 2018 and from 2018 to 2019 he was a lecturer at the department of computer engineering, Kalasin University, Thailand, and the department of teacher training in electrical engineering, RMUTI, Thailand, respectively. From September 2019 until the present, he has been a member of the research center for quantum technology, Faculty of Science, Chiang Mai University, as a staff research scientist. His scientific interests mainly focus on modeling and simulation of several materials, including magnetic insulators, topological insulators, magnetic topological material, antiferromagnetic topological materials, and 2D (Janus) magnetic materials, using a combination of density functional theory (DFT), quantum transport based on non-equilibrium Green function (NEGF), atomistic spin modeling, spin Hamiltonian, and finite element.

 micromachines

MDPI

Editorial

Editorial for the Special Issue on Magnetic and Spin Devices

Viktor Sverdlov [1,*] and Nuttachai Jutong [2,*]

[1] Christian Doppler Laboratory for Nonvolatile Magnetoresistive Memory and Logic, Institute for Microelectronics, TU Wien, 1040 Vienna, Austria
[2] Research Center for Quantum Technology, Faculty of Science, Chiang Mai University, Chiang Mai 50200, Thailand
* Correspondence: sverdlov@iue.tuwien.ac.at (V.S.); nuttachai.jutong@cmu.ac.th (N.J.)

As scaling of semiconductor devices displays signs of saturation, the focus of research in microelectronics shifts towards finding new computing paradigms based on novel physical principles. The electron spin is another intrinsic characteristic of an electron offering additional functionality to the electron charge-based semiconductor devices currently employed in microelectronics. Several fundamental issues including spin current injection, spin propagation, and relaxation, as well as spin orientation manipulation by the gate, have successfully been resolved enabling the electron spin for digital applications.

In order to generate and detect the spin-polarized currents by electrical means, magnetic metal contacts can be employed. The ferromagnetic contacts discussed by Boroš et al. [1,2] should be sufficiently small to comprise a single magnetic domain with a clear magnetization orientation. Magnetic moments of small domains were successfully used in the past and are still employed to store the information in magnetic hard disc drives. Herewith, the binary information is encoded into the domain's magnetization orientation.

The magnetization of a domain creates a stray magnetic field which can be detected. Alternating magnetic moments create magnetic fields in opposite directions. Read heads can detect the field and read the information. It is shown by Khunkitti et al. [3] that magnetic heads with high sensitivity are important ingredients to enable ultrahigh areal densities of magnetic data storage technology. To write information, a magnetic field close to the magnetic domain is created by means of the electric currents running into the magnetic head. The recording density critically depends on the magnetic media properties as illustrated by Khunkitti et al. [4].

Without an external magnetic field, the magnetization of a magnetic domain is preserved and does not change over time. Therefore, adding magnetic domains in electronic devices introduces nonvolatility—the ability to preserve the functional state of a device without external power supplied. In addition, one can manipulate the magnetization orientation of a small magnetic domain by running a spin-polarized current through it. If the current is sufficiently strong, the magnetization of the domain aligns parallel with the spin current polarization. A purely electrical manipulation of a magnetic domain by electron currents provides an exciting opportunity to develop a conceptually new type of nonvolatile memories with improved scalability. The impinging spin-polarized current can be created by the charge current running through another ferromagnet separated from the small domain by a metallic spacer or a tunnel barrier. The electrical resistance of a sandwich made of two ferromagnetic contacts separated by a tunnel barrier (or a spacer) strongly depends on the relative magnetization orientation of the contacts in either parallel or anti-parallel configuration. Therefore, the binary information encoded into the relative magnetization is revealed through the sandwich's resistance. This type of emerging memory is termed magnetoresistive. Magnetoresistive memories possess a simple structure. They offer excellent endurance and high speed of operation. Magnetoresistive memories are compatible with the metal–oxide–semiconductor field-effect transistor fabrication process. They open perspectives for conceptually new low power computing paradigms of data

Citation: Sverdlov, V.; Jutong, N. Editorial for the Special Issue on Magnetic and Spin Devices. *Micromachines* **2022**, *13*, 493. https://doi.org/10.3390/mi13040493

Received: 18 March 2022
Accepted: 19 March 2022
Published: 22 March 2022

processing in the nonvolatile segment with the same devices used to store and to process the information.

Alternatively, spin currents can be generated by the spin Hall effect, when a pure spin current running normal to the interface of a heavy metal slab is produced by the charge current running through the slab. A torque due to the spin current acts on a ferromagnetic layer positioned on top of the heavy metal slab and may change the layer's magnetization orientation. The size of the ferromagnetic layer determining the dimension of the memory cell can be significantly reduced if the magnetic layer is polarized perpendicularly to the interface of the ferromagnet layer with the slab; however, special care must be taken to deterministically switch the magnetization without an external magnetic field [5,6].

In line with most advanced spintronic devices, magnetic properties of ferromagnets are also exploited in a more traditional way as magnetic contacts in various security applications, ranging from access allowing magnetic cards to detectors in electrical security systems where the magnetic parts are installed on windows, door, or other objects whose change of state triggers an alarm as discussed by Boroš et al. [1]. The reliability parameters of the magnetic contact must be preserved within 10% variations regardless of extreme conditions of their functionality, which requires creation of a test environment allowing to stress the contacts under extreme conditions. Special experiments were designed and performed by Boroš et al. [1] in an air chamber with the temperature changing from -70 °C to $+180$ °C to investigate relations between the effect of temperature and the closing/opening distances for five selected types of magnetic contacts intended for outdoor use in. The automation of the testing process is essential to accelerate the reliability benchmarking and to eliminate human error. For this reason, the testing infrastructure described by Boroš et al. [2] was designed to use internal memory for evaluating the number of successful closures of magnetic contacts and then transmit the digitized data.

Current-perpendicular-to-the-plane giant magnetoresistance (CPP-GMR) sensors are promising as read heads for ultrahigh (> 1 Tb/in^2) density magnetic data storages. The dependence of the free layer thickness on the stability of the $Co_2(Mn_{0.6}Fe_{0.4})Ge$ Heusler-based CPP-GMR sensors was investigated by Khunkitti et al. [3]. The magnetoresistance ratio (MR), readback signal, dibit response asymmetry parameter, and power spectral density were calculated and characterized. It was demonstrated that the read head with a free layer thickness of 3 nm displays the best stability with a large MR of 26%.

To boost the density of magnetic storages even further, the magnetic media, as well as the data recording, must be improved. Exchange-coupled-composite-bit-patterned media (ECC-BPM) with microwave-assisted magnetic recording to improve the writability of the magnetic media at a 4 Tb/in^2 recording density was proposed by Khunkitti et al. [4]. It was demonstrated that the switching field of the bilayer ECC-BPM is significantly reduced, with the lowest value of 12.2 kOe achieved at 5 GHz microwave frequency.

Spin-orbit torque magnetoresistive random access memory combines high-speed access with excellent endurance and is promising for application in caches. One of the options to achieve deterministic switching of a free layer with perpendicular magnetization without an external magnetic field is to use a two-pulse dynamic approach: the first pulse pre-selects the cell and puts the magnetization in-plane, while the second pulse completes the switching. As the switching probability depends on the amplitude and duration of the current pulses, finding an optimal sequence is paramount. A reinforcement learning approach in combination with micromagnetic simulations is introduced by de Orio et al. [5] to optimize the switching of a spin-orbit torque memory cell. The approach allows not only to find an optimal pulse sequence, but also to analyze the impact of material parameter variations on deterministic switching.

Field-free switching in perpendicular magnetic tunnel junctions is achieved by combing the spin-transfer (ST) and spin-orbit (SO) torques as demonstrated by Wasef and Fariborzi [6]. In particular, the relationship between the ST and SO critical current densities is obtained. The relationship between the critical SO and ST current densities depend not only on damping, but also on the magnitudes of the field-like torques opening the path towards designing magnetoresistive memories operated with help of both SO and STT currents.

In conclusion, we would like to thank all the authors for submitting their papers to this Special Issue. We also thank all the reviewers for dedicating their time and helping to improve the quality of the submitted papers. Last but not least, we would like to thank Ms. Aria Zeng from the *Micromachines* publishing office for her constant guidance and great support.

Conflicts of Interest: The authors declare no conflict of interest.

References

1. Boroš, M.; Velas, A.; Soltes, V.; Dworzecki, J. Influence of the Environment on the Reliability of Security Magnetic Contacts. *Micromachines* **2021**, *12*, 401. [CrossRef] [PubMed]
2. Boroš, M.; Velas, A.; Zvaková, Z.; Šoltés, V. New Possibilities for Testing the Service Life of Magnetic Contacts. *Micromachines* **2021**, *12*, 479. [CrossRef] [PubMed]
3. Khunkitti, P.; Siritaratiwat, A.; Pituso, K. Free Layer Thickness Dependence of the Stability in Co2(Mn0.6Fe0.4)Ge Heusler Based CPP-GMR Read Sensor for Areal Density of 1 Tb/in^2. *Micromachines* **2021**, *12*, 1010. [CrossRef] [PubMed]
4. Khunkitti, P.; Naruemon Wannawong, N.; Jongjaihan, C.; Siritaratiwat, A.; Kruesubthaworn, A.; Kaewrawang, A. Micromagnetic Simulation of L_{10}-FePt-Based Exchange-Coupled-Composite-Bit-Patterned Media with Microwave-Assisted Magnetic Recording at Ultrahigh Areal Density. *Micromachines* **2021**, *12*, 1264. [CrossRef] [PubMed]
5. De Orio, R.L.; Ender, J.; Fiorentini, S.; Goes, W.; Selberherr, S.; Sverdlov, V. Optimization of a Spin-Orbit Torque Switching Scheme Based on Micromagnetic Simulations and Reinforcement Learning. *Micromachines* **2021**, *12*, 443. [CrossRef] [PubMed]
6. Wasef, S.; Fariborzi, H. Theoretical Study of Field-Free Switching in PMA-MTJ Using Combined Injection of STT and SOT Currents. *Micromachines* **2021**, *12*, 1345. [CrossRef] [PubMed]

 micromachines

Article

Influence of the Environment on the Reliability of Security Magnetic Contacts

Martin Boros [1,*], Andrej Velas [1], Viktor Soltes [1] and Jacek Dworzecki [2]

1 Department of Security Management, Faculty of Security Engineering, University of Zilina, Univerzitna 8215/1, 010 26 Zilina, Slovakia; andrej.velas@fbi.uniza.sk (A.V.); viktor.soltes@fbi.uniza.sk (V.S.)
2 Faculty of Security Science, Military University of the Land Forces in Wroclaw, Piotra Czajkowskiego 109, 51-147 Wrocław, Poland; jacekdworzecki@o2.pl
* Correspondence: martin.boros@fbi.uniza.sk; Tel.: +421-41-513-6668

Abstract: Magnetic contacts are one of the basic components of an alarm system, providing access to buildings, especially windows and doors. From long-term reliability tests, it can be concluded that magnetic contacts show sufficient reliability. Due to global warming, we can measure high as well as low ambient temperatures in the vicinity of magnetic contacts, which can directly affect their reliability. As part of partial tests, research into the reliability of magnetic contacts, we created a test device with which their reaction distance was examined under extreme conditions simulated in a thermal chamber. The results of the practical tests have yielded surprising results.

Keywords: magnetic contacts; reliability; practical tests; reaction distance; extreme conditions

Citation: Boros, M.; Velas, A.; Soltes, V.; Dworzecki, J. Influence of the Environment on the Reliability of Security Magnetic Contacts. *Micromachines* **2021**, *12*, 401. https://doi.org/10.3390/mi12040401

Academic Editor: Viktor Sverdlov

Received: 20 February 2021
Accepted: 3 April 2021
Published: 5 April 2021

Publisher's Note: MDPI stays neutral with regard to jurisdictional claims in published maps and institutional affiliations.

1. Introduction

Magnetic contacts and magnetic fields have become an integral part of our lives, even though most of the time we would not even realize it. Thanks to our advantageous capabilities, we have been using them for almost a hundred years. They were developed in 1936 in the laboratories of Bell Telephone Laboratories by Walter B. Elwood, and three years later, in 1940, the first patent record was recorded, which is almost identical to the magnetic contacts used today [1]. With the development of time, the possibilities of using magnetic contacts in various branches of industry gradually developed, from compasses to handling equipment, computer memories, hard disks, storages, payment cards, and many other possibilities. In addition to magnetic contacts, formed by a permanent magnet and a reed contact, magnetic loops are also very often used in practice. Within the scientific community, the magnet or its components have been used by several scientists, such as Macholí Belenguer and his team [2], which have explored the possibilities of identifying vehicles by changing the voltage in the magnetic loops. On the other hand, the authors Primin and Nedayvoda [3] focused their research on the possibilities of using magnetic fields in biological identification using a SQUID sensor, while Ponce et al. [4] dealt with the efficiency of using magnetic sensors in water meters. However, these and other important scientific works suggest that the importance of magnetic contacts in science is very important.

The security sector is no exception, in which several possibilities of magnetic contact use are available, such as in the case of control of inputs in which magnetic cards are used or as detectors in an electrical security system [5,6]. It is the latter option that is one of the basic options for securing entrance openings within the building envelope, such as doors or windows. Thanks to their characteristics and the possibility of use, they are also used in perimeter protection (securing the entrance/exit gate from the premises) or also object protection. As mentioned, the base of the magnetic contacts is formed by a permanent magnet and a tongue, the ferromagnetic contacts of which are encased in a thin glass tube with a protective atmosphere to prevent corrosion. The magnetic part is usually installed on a moving part such as a window, door, or other objects whose change of state triggers an alarm. The tongue or contact is mounted on a fixed part, usually a door frame. From a functional point of view, we then divide the magnetic contacts into normally open (NO) and normally closed (NC) [1,6,7].

The NO magnetic contacts close when the switching magnet approaches, allowing the passage of electrical current. On the contrary, the NC contacts are closed without the action of a magnetic field and allow the passage of an electric current. This is because they contain another magnet. The closing of the magnets is caused by the addition of another magnet, i.e., another permanent magnet is added to the NO type [1,8]. The difference in the evaluation of the magnetic contacts is shown in Figure 1.

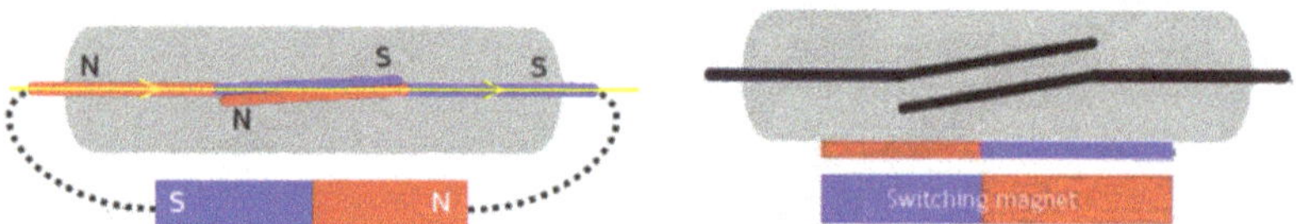

Figure 1. Example of making magnetic contacts on the left type of normally open (NO), on the right type of normally closed (NC).

The actual spacing of the contact parts is only a few microns, so these contacts are very sensitive. The contacts are manufactured in different versions for different security levels and environmental classes. We know magnetic contacts in metal and plastic housings as well as contacts intended for surface mounting or drilling mounting. A special group is the magnetic contacts connected to the control panel wirelessly. Magnetic contacts can be overcome in several ways, especially by using a parasitic magnet with a high magnetic field strength or by bridging the cabling. For these cases, sabotage circuits consisting of differently oriented magnetic contacts are also inserted into magnetic contacts intended for higher security classes [1,8,9].

As mentioned above, magnetic contacts represent one of the basic types of alarm system detectors, and therefore all the same requirements apply to them as to other elements of the system. It is usually a matter of defining the degree of security and the environment class, according to the European technical standard EN 50131-1 and the separate functional requirements comprehensively stated in the European technical standard EN 50131-2-6 determined by magnetic contact [9–11]. Security levels, according to EN 50131-1, are divided into four categories with a designation from 1 (lowest risk) to 4 (high risk). The four types are also divided into categories, environmental classes, the first two of which are intended for indoor and the second two for external environments [10]. The user should follow the manufacturer's recommendations and use the detectors only in the recommended environment class and at the required security level. Failure to correctly classify the environment may result in a malfunction of the detector, and thus of the electrical system as a whole [6]. According to [11], the reliability parameters of the magnetic contact must be met to a tolerance of $+/-10\%$, unless exceptions are defined. The mentioned standard defines several possibilities of performing functional tests of magnetic contacts to determine the correct functionality from the point of view of electrical requirements, switching requirements, detection requirements. A separate group of tests of magnetic contacts—as well as other detectors of intruder alarm systems—consists of environmental tests, which are addressed in the European technical standard EN 50130-5 [11,12].

The need for reliability testing in extreme ambient conditions is significant mainly due to global warming, in which we have witnessed extreme fluctuations in temperature values in the mid-range in recent years. In recent years, the number of research focusing on extreme weather conditions has also increased, as people realize the need to test different types of systems under ambient loads. For example, in the conditions of the Slovak Republic, the deviation of average temperatures recorded in 2018 increased by 1.3 °C compared to previous years, which corresponds to the long-term average warming of

Europe and the whole planet [13]. However, due to global pollination, we also encounter the so-called "Arctic winters", when the temperature drops well below freezing. In the coldest village in Slovakia, Oravská Lesná, the lowest temperature in the last decade was measured in 2017 at −35.5 °C. In the long run, it is, therefore, appropriate to consider whether the definition of four environmental classes with a specific temperature range is sufficient, as in the highest class, class IV, a temperature range from −25 °C to +60 °C is calculated [10,12,13].

To investigate the influence of temperature on the magnetic characteristics of NdFeB permanent magnets, the authors Calin and Helerea investigated the influence of temperature on the demagnetization curves of cylindrical magnets in the temperature range from 25 °C to 120 °C. According to their findings, the properties of cylindrical magnets change at approximately 50 °C [14]. Our research aims to express the working distance of magnetic contacts within which it can detect the intruder even in degraded ambient conditions. We believe that despite the results of the mentioned research, the working distance would not have to be so affected. Within our assumptions, we are working with the knowledge that the magnetic properties are influenced not only by temperature but also by other influencing parameters such as the shape and material of the magnet or the surrounding metal objects. We also base our knowledge on the average, maximum operating temperature of magnetic contacts being 80 °C.

Due to extreme weather conditions, the design of a resilient distribution network was addressed by Shahbazi A and colleagues, who achieved very good system resilience results in their case studies [15]. Another interesting and significant research was conducted by Bennett, JA, and colleagues, who focused on modeling the optimization of energy system architecture as prevention against hurricane impacts [16]. We decided to research the influence of extreme weather influences on the reliability of magnetic contacts as one of the partial parts of comprehensive research into the components of electrical security systems. We did not rely on the conditions and recommendations of European technical standards, as we wanted to simulate real conditions where the perpetrator may not follow the procedures prescribed by the standard. We followed up on the research activities of the workplace, within which we focused on the evaluation of camera systems against extreme weather conditions, as well as many other important tests of transmission systems, communication interface, location system, and the like [16–20].

Experimental testing of extreme weather effects on camera systems has shown the fact that they can fully operate outside the temperature ranges specified in the technical standard. It was found that a camera system designed for the second class of environment, −10 °C to +40 °C, is able to fully operate at a temperature of +120 °C [13]. The problem, in this case, occurred at a value of +160 °C, when the system stopped and appeared to be without a camera connected. However, after sufficient cooling, the camera continued to function but showed permanent image damage due to sensor damage [21]. Based on the findings, we also decided to further verify the functionality of the security system components. Comprehensive and extensive findings could be used as a basis for adjusting the environmental classes specified in the technical standard for security systems. If further component testing is possible, the magnetic contacts should be used, as this is the most commonly used component of the security system, designed to protect the opening parts in the building or the premises of the building. Another reason for choosing magnetic contacts was their financial simplicity and the fact that they are the most important element, especially in the case of entering the building, which must function properly for the correct detection of the intruder. From the tests of camera systems, it is possible to use and focus only on the realistically achievable thermal values, while the limit temperature of the functionality of the magnet should not be exceeded. In the case of a camera system, such values were not and are not generally known, and therefore we went to huge extremes, but at the cost of long-term damage to the camera [10,13,21].

2. Materials and Methods

As part of the evaluation of the impact of extreme conditions, we decided to perform experimental measurements using our test equipment in the air conditioning chamber. Using experimental measurements, we investigated the causality between the effect of temperature and the closing/opening distances of five selected types of magnetic contacts intended for outdoor use.

The basic pillar of measurements was the mentioned air-conditioning chamber, brand VÖTSCH VCL 7010, which we have available within the material and technical equipment of the laboratory. The air conditioning chamber provides a temperature range from −70 °C to +180 °C and a heating and cooling rate of 3.5 K/min [22].

We created a test device especially for this type of measurement, which consisted of two hardboard plates with a length of 50 cm and a width of 3 cm. At the ends of both boards, we attached two wooden prisms with glue, which were used to hold the tested magnetic contacts, and we placed the boards on top of each other. The model of the upper part of the test device was created in the Sketch UP! program, as shown in Figure 2. We decided to use wooden prisms, as they are a non-conductive material, and so will not affect the detection and switching functionality of the tested magnetic contacts. We also used glue rather than iron screws to attach them. For better handling of the test equipment, we connected the boards in several places with mounting tapes to perform only one direction of movement [23].

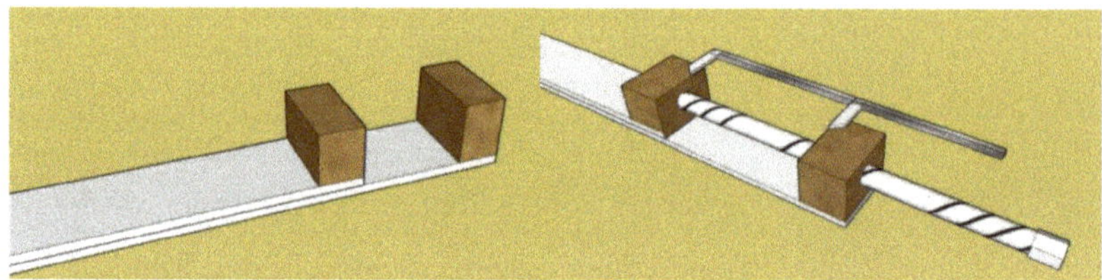

Figure 2. Graphic design of the test device from the top right, the drive part from the left.

At the other end of the boards, we placed a drive, which was formed by a threaded rod with a diameter of 8 mm. With the help of one turn of the threaded rod, it was possible to move the wooden prisms by 1.25 mm, which achieved a very precise and sensitive mechanical movement. We achieved the drive-by drilling wooden prisms and then inserting a threaded screw into a hole on a fixed (bottom) plate. To achieve, in addition to movement, the measurement of the closing distance of the tested magnetic contacts, we also attached a caliper to the wooden prisms, with the help of which we were able to express the given distance. The model of the lower part of the test device together with the drive, created in the program Sketch UP! is shown in Figure 2. The test device in the air conditioning chamber is shown in Figure 3.

We read the values manually from the caliper, so it was possible to assume that a measurement error had occurred. We rounded the values to 0.5 mm, which could cause a measurement error of ±0.25 mm. We designed the test device to be able to zoom in and out on the magnetic contacts in the air conditioning chamber. We adapted its size to the size of the chamber opening. We used an optical and acoustic indicator to determine the activity/inactivity of the magnetic contact. The indicator consisted of LEDs, piezoelectric buzzer with a power supply. We attached the indicator constructed in this way to the magnetic contact, and in the case of its switching on, i.e., activation, the LED lit up and the buzzer sounded. The figure of the indicator, as well as its wiring diagram, are shown in Figure 4.

Figure 3. Test equipment placed in an air-conditioning chamber during experimental testing.

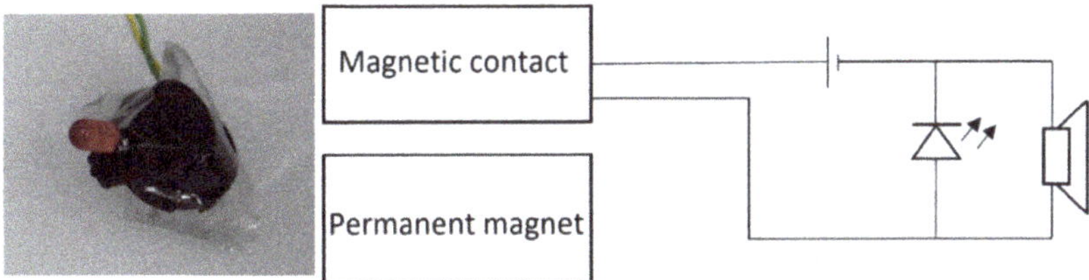

Figure 4. Indication device (right) and its wiring diagram.

As mentioned in total, we tested five different magnetic contacts in experimental tests and followed the following procedure for each measurement:

- Study of the enclosed technical sheet of the magnetic contact from the manufacturer,
- Fixing the magnetic contact to the test equipment, usually by double-sided adhesive tape or by impregnation into holes in wooden prisms,
- Implementation of experimental tests:
  - measuring the closing distance—movement from 0 mm to 100 mm,
  - measuring the opening distance—movement from 100 mm to 0 mm,
- Evaluation of the measurement process and achieved results.

We measured the temperature range from −40 °C to +70 °C, while at negative values we changed the temperature by 5 °C and in the case of positive values by 10 °C. For each heat value, we performed two types of measurements, switching on and off, according to the above procedure. We stopped for two minutes between the individual distance measurements to prevent the measurement from being devalued due to the lasting closing/opening action from the previous measurement. We proceeded in descending order from positive to negative values to avoid possible corrosion. We recorded the measured values and expressed the average values using the average function in MS Excel. Subsequently, we used the STDEV function in MS Excel to express the standard deviation. In the case of each temperature change, we allowed the magnetic contact to acclimatize for 10 min, and then we performed 10 repetitions of the magnetic contact closing measurement. The transition between temperatures was relatively fast, falling for tens of minutes. The values given in the results section represent the arithmetic mean of the measured values for the individual temperatures. For those magnetic contacts for which the closing and opening did not occur at the same distance, we also determined the hysteresis distance by the difference between the closing and opening distance.

3. Results

We were the first to test the magnetic contact with the marked MAS 203 from the Czech manufacturer Asita. In this case, the manufacturer does not state the values of closing and opening of the magnetic contact, but only the working distance of 0–30 mm with a tolerance of 2 mm within which it should work fully. The installation requirements also state that the parallelism of the installation of the magnetic contacts must be observed. To ensure correct installation, we used double-sided adhesive tape, the anchoring of the magnetic contact during the measurement on the test device is shown in Figure 5A. The results obtained in the measurement are shown in Table 1.

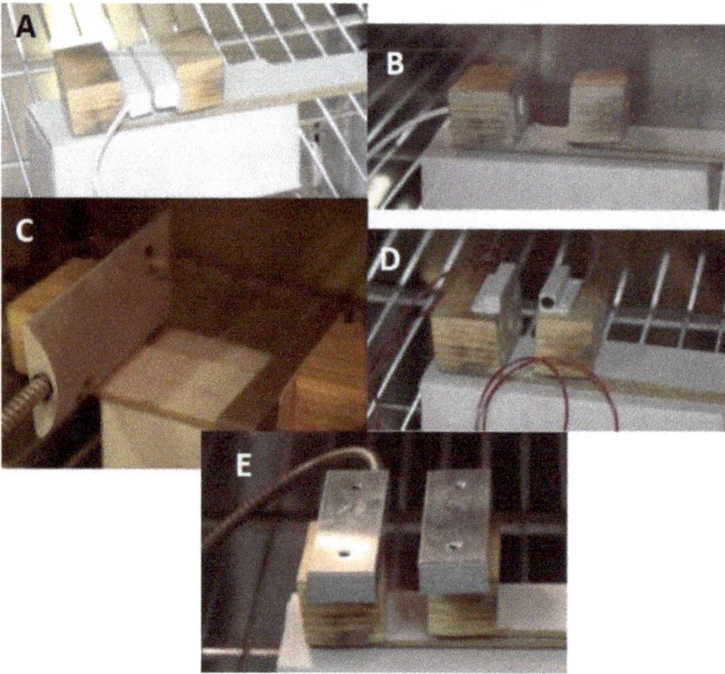

Figure 5. Magnetic contacts during experimental testing, (**A**)-MAS 203, (**B**)-MAS 333, (**C**)-SA 220, (**D**)-USP 131, (**E**)-USP 500.

Table 1. MAS 203 magnetic contact on and off values during experimental tests.

Temperature [°C]	Clamping Distance [mm]	Standard Deviation	Opening Distance [mm]	Standard Deviation
−40	28.5	1.96	28.5	1.74
−35	29	1.93	29	1.08
−30	29.5	0.15	29.5	0.23
−25	30	1.97	30	1.79
−20	31	0.35	31	1.62
−15	32	0.45	32	0.34
−10	33	1.41	33	1.62
−5	34	0.91	34	0.35
0	34	0.45	34	0.87
10	34.5	1.41	34.5	0.30
20	35	0.41	35	0.63
30	35	0.22	35	0.91
40	34	0.34	34	1.86
50	34	1.10	34	0.20
60	34	0.87	34	0.21
70	34	1.21	34	1.24

As we can see due to the higher temperature, in particular, the working distance of the magnetic contact increased by 2–3 mm compared to the values given by the manufacturer, even with the permissible tolerance. Even for temperatures of 0 °C or temperatures around 20 °C (standard outdoor temperature), the distance was covered even when considering tolerance. On the contrary, negative values slightly reduced the working distance values. According to experimental tests of the reliability of a given magnetic contact, the average working distance was at the level of 30.5 mm [24]. By measuring, we found that the magnetic contact does not work with any opening hysteresis, i.e., the closing and opening of the magnetic contact occurs at the same distance [25]. Throughout the measurement, we followed the procedure created and determined by us, listed in Section 2.

The second magnetic contact tested was from the same manufacturer but with the type designation MAS 333. According to the information provided by the manufacturer, this type of magnetic contact can be used in environment-class III, i.e., in the temperature range from −25 °C to +50 °C. Since we followed the same measurement procedure for all magnetic contacts, we performed measurements at temperatures from −40 °C to +70 °C and these results are shown in Table 2. The working distance of the magnetic contact MAS 333 is 0–22 mm with tolerance according to the manufacturer 2 mm. As this is a countersunk type of magnetic contact, we used holes in the wooden prisms of the test device to attach it [26]. The location of the magnetic contact in the test device during the measurement is shown in Figure 5B.

Table 2. MAS 333 magnetic contact on and off values during experimental tests.

Temperature (°C)	Clamping Distance (mm)	Standard Deviation	Opening Distance (mm)	Standard Deviation
−40	22	0.53	23.5	0.33
−35	22	0.43	24	0.30
−30	22.5	1.78	24	1.86
−25	23	0.84	25	0.79
−20	23	1.18	26	0.87
−15	23	1.15	26	0.42
−10	23	1.19	25	0.25
−5	23	0.34	25	0.58
0	23	0.29	25	1.86
10	23.5	1.83	25	0.75
20	23	0.47	25	1.69
30	22.5	1.59	25	1.97
40	22.5	1.20	24.5	1.95
50	22	1.92	24.5	0.66
60	22	1.49	24	0.43
70	22	1.69	24.5	0.46

From the achieved results we can state that the magnetic contact in case of closing always worked in the range of the working distance specified by the manufacturer, taking into account also the allowed tolerance. Conversely, in the case of opening the activation of the magnetic contact, we measured different values, exceeding the working distance, in the temperature range from −25 °C to +70 °C. The different values were caused by a hysteresis phenomenon, the values of which are shown in Figure 6 [23].

Another tested magnetic contact had the type of designation SA220 and is primarily offered by Jablotron in its portfolio. According to the information from the manufacturer, the magnetic contact is intended for environment-class IV, i.e., the highest class, with a temperature range from −25 °C to +60 °C. The body of the magnetic contact, the switching part, is housed in a metal housing and the permanent magnet in a plastic housing. Dimensionally, this is the largest magnetic contact with atypical dimensions and omnidirectional radiation, i.e., according to which axis within the three-dimensional space is installed, the distance of switching on and off is determined. As part of the measurements, we decided to use an installation with a z-axis, as due to its size, the manufacturer states the largest values of clamping (64 mm) and expansion (77 mm). We used double-sided adhesive tape for installation on the test equipment, the magnetic contact during the measurement is shown in Figure 5C. Even in this case, we worked with a temperature range from −40 °C to +70 °C.

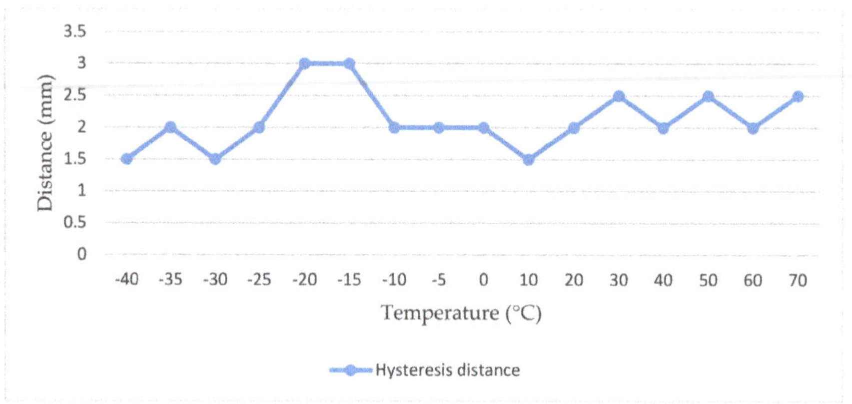

Figure 6. Graphical representation of the values of the hysteresis phenomenon of the magnetic contact MAS333.

As we can see in Table 3, we observe an almost exponential dependence of distance on temperature. While at negative values, the closing and opening distances were several tens of millimeters below the level specified by the manufacturer. Conversely, in the case of high temperatures, the values of switching on and off were much higher. Compliance with the data given by the manufacturer was achieved only in the temperature range −5 °C to 0 °C when closing and +10 °C to 20 °C when opening the magnetic contact. For measurement, we also recorded a significantly high value of hysteresis distance, in some cases at the level of up to 6 mm. The complete expression of the hysteresis distance is shown in Figure 7 [23,24,26]. The measurement results are quite worrying, as the magnetic contact did not work properly in the temperature range in which it was recommended to be installed.

Table 3. SA 220 magnetic contact on and off values during experimental tests.

Temperature (°C)	Clamping Distance (mm)	Standard Deviation	Opening Distance (mm)	Standard Deviation
−40	54	3.35	57	3.39
−35	54.5	2.78	57.5	4.16
−30	54.5	3.77	58.5	2.45
−25	56.5	2.53	59	4.94
−20	58	4.03	60.5	2.74
−15	59	3.45	62	4.83
−10	61	4.87	64	2.62
−5	63.5	3.16	66	3.07
0	66	4.50	69	2.91
10	72	3.94	76	4.94
20	81.5	3.90	86	4.91
30	84	4.69	89	3.01
40	85	2.03	90.5	3.95
50	87	3.34	92	3.50
60	88	4.54	94	3.27
70	89	2.70	94	2.02

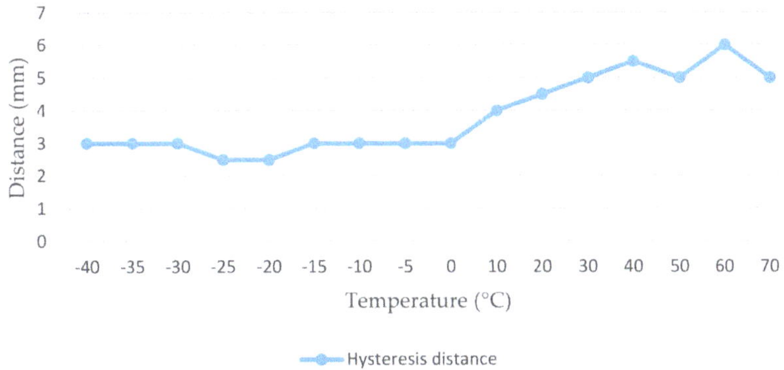

Figure 7. Graphical representation of the values of the hysteresis phenomenon of the magnetic contact SA 220.

Another magnetic contact tested came from United Security Products and was type-named USP 131. It is a small magnetic contact, almost 35 mm, housed in a plastic case. According to the manufacturer, the magnetic contact can work with a working distance of 20 mm, while it is designed for environment-class III, i.e., within the temperature range from −25 °C to +50 °C. Due to the uniformity and adherence to the established methodological procedure, we subjected the magnetic contact to reliability testing in the temperature range of −40 °C to +70 °C. The results obtained in the measurements are given in Table 4. The magnetic contact, USP 131, has an elongated part on the lower part of both components intended for their assembly, so we decided to place it on top of the test device. The location of the magnetic contact during the measurement is shown in Figure 5D.

Table 4. USP 131 magnetic contact on and off values during experimental tests.

Temperature (°C)	Clamping Distance (mm)	Standard Deviation	Opening Distance (mm)	Standard Deviation
−40	21	1.11	16	1.09
−35	21	1.13	16	0.55
−30	21	0.91	16	0.60
−25	21	1.75	16.5	0.63
−20	21	0.87	16.5	1.48
−15	21	1.92	16.5	1.40
−10	21	0.63	16.5	0.66
−5	21	1.45	16.5	2.00
0	21.5	0.39	16.5	1.34
10	22	1.91	17	0.47
20	21.5	1.16	17	1.45
30	22	1.85	17	1.84
40	22	1.94	16.5	1.96
50	21.5	1.46	16	1.97
60	22	1.36	16	0.99
70	22	0.32	16	1.19

As we can see from the achieved results, the manufacturer's stated working distance of 20 mm was not reached in any of the measured temperatures. A positive finding is that the closing and opening distance of the magnetic contact was almost constant. Another positive is that the magnitude of the hysteresis distance that we observed when switching the magnetic contact is also, almost at a constant level, gradually increasing from a temperature of 30 °C [9,10]. The hysteresis distance is shown in Figure 8.

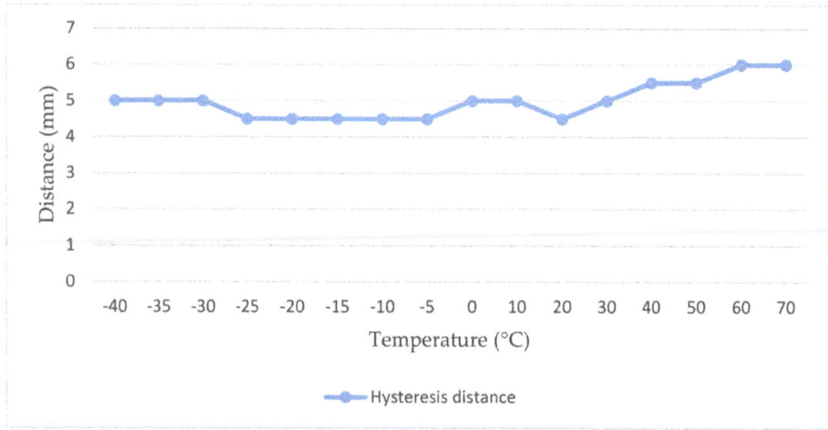

Figure 8. Graphical representation of the values of the hysteresis phenomenon of the magnetic contact USP 131.

The last magnetic contact tested had the designation USP 500 and was therefore from the same American company as the previous contact. It is a relatively durable-looking magnetic contact designed for surface installation as it is formed by two iron, elongated blocks. In this case, too, it was a magnetic contact intended for environment-class III and was subjected to measurements at the same temperature levels as all other magnetic contacts. The working distance specified by the manufacturer is 63.5 mm, the attachment of the magnetic contacts during the measurement was realized with the help of double-sided tape and is shown in Figure 5E.

As we can see from the results in Table 5, the magnetic contact did not even approach the working distance given by the manufacturer at sub-zero temperatures, at −40 °C, there was even a difference of almost 20 mm. An indication of the approach to the data from the manufacturer can be observed only at a temperature of 40 °C, but even in this case, the difference between the distances was at the level of 3 mm. When the switching element is switched on and off by a permanent magnet, it was possible to observe a hysteresis phenomenon of small size when compared to some of the other magnetic contacts. The values of the hysteresis phenomenon are shown in the graphic design in Figure 9.

Table 5. USP 500 magnetic contact on and off values during experimental tests.

Temperature [°C]	Clamping Distance [mm]	Standard Deviation	Opening Distance [mm]	Standard Deviation
−40	47.5	2.53	48.8	2.03
−35	48.5	2.37	49.5	2.69
−30	50.5	3.63	52	3.34
−25	51.5	3.26	52.5	2.03
−20	52.5	2.73	54	2.13
−15	53.5	2.73	55	3.75
−10	54.5	2.57	56	2.39
−5	55.5	2.54	56.5	2.79
0	56.5	3.76	58	3.72
10	55.5	2.08	57	2.15
20	56.5	3.12	58	2.20
30	57.5	2.52	59.5	3.45
40	58.5	2.03	60.5	2.67
50	58.5	2.55	60.5	3.49
60	58.5	2.17	60	2.92
70	57.5	3.72	59.5	2.21

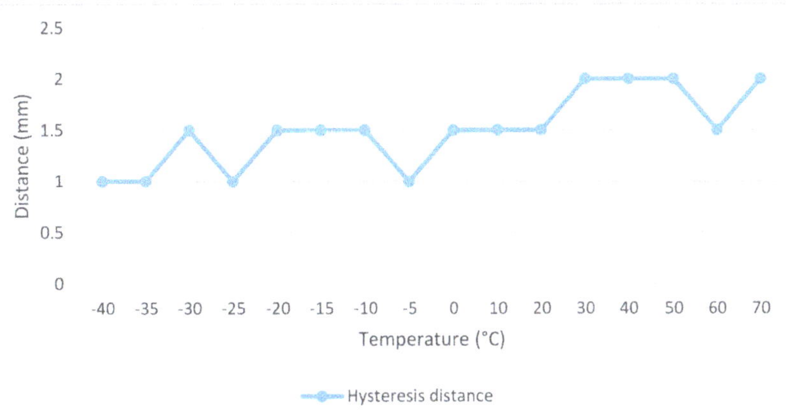

Figure 9. Graphical representation of the values of the hysteresis phenomenon of the magnetic contact USP 500.

4. Discussion

Through experimental tests that we performed in the air conditioning chamber, we can conclude that several of our assumptions have been confirmed. The reliability of the safety magnetic contacts is influenced by the shape of the magnetic contacts. We base this statement on the results in which we found that in the case of the four tested magnetic contacts, we measured relatively reliable data, within which the magnetic contacts would be able to function as expected. In one case, SA220, there was an appropriate difference as the measured values were almost 20 mm different than stated by the manufacturer. The possibility that the manufacturer incorrectly determined in the technical file the position of the sides of the magnet within the x, y, z axes would also be considered. However, even if we consider that we connected the magnetic contact oppositely, in both other possibilities the magnetic contact showed a difference of at least 15 mm in size. Based on these results, we must conclude that the magnetic contact is unreliable, even in the environment specified by the manufacturer.

In other cases of measurement, we achieved interesting results, in which we achieved almost identical results in all tested temperatures. By default, we achieved a difference of approximately 2 mm between negative and positive temperatures. It was confirmed that the cold benefits the functionality of the magnetic contacts, as we measured values similar to those stated by the manufacturer at negative values. Based on these results, we can conclude that the other four magnetic contacts are reliable despite the lower, measured distances. Most of the measured values were within the working distance specified by the manufacturer or within the 10% tolerance allowed by the technical standard [11]. We achieved interesting and surprising results in positive and high temperatures, respectively, where there was no rapid change in working distance as expected. It can be stated that we correctly assumed that within the standard operating temperature of the magnetic contacts, the shape of the block, the magnetic induction does not change to such an extent as to affect the reliability of the safety magnetic contacts.

As they are undoubtedly interesting and new findings, we decided to perform a simple measurement to determine the strength of the permanent magnet, which is used to switch the magnetic contact. For the experiment, we used a test device created by us and a steel cylinder made of 18 washers, a screw, and a butterfly nut. The measurement consisted of setting the initial position, which was seven centimeters. We placed a permanent magnet on the sliding part of the test device and the mentioned steel cylinder on the fixed part, always so that the edge of the cylinder is aligned with the edge of the wooden prism. Gradually, we moved the test device until the moment when the steel cylinder, due to the attractive force of the magnet, did not start moving towards the permanent magnet. The block diagram of the measurement is shown in Figure 10. We performed the measurement at several temperatures and with twenty repetitions, the results of the measurement are shown in Table 6.

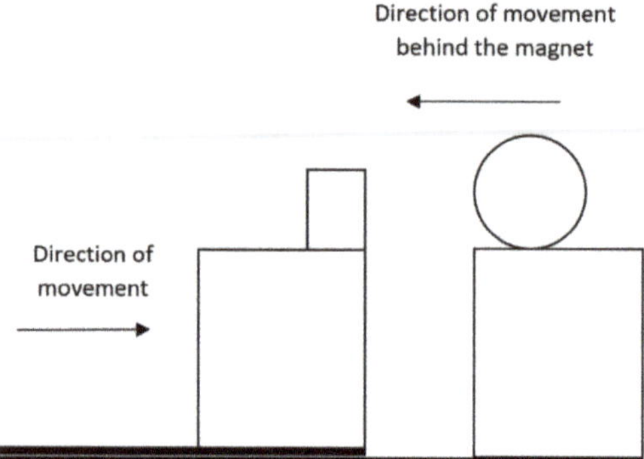

Figure 10. Block diagram of magnetic field strength measurement.

Table 6. Measured values of magnetic field strength.

Temperature	−10	−5	0	15	25	35
Arithmetic mean of measured values [mm]	11.91	11.92	11.92	11.92	11.92	11.92
Standard deviation	0.09	0.07	0.08	0.07	0.08	0.07

As can be seen from the measured values, the attractive force of the magnetic field does not change under the influence of temperature—or rather, changes to a negligible extent. For this simple measurement, we used a magnetic contact USP 131. The reduced force values, compared to Table 4, are caused by the weight of the steel cylinder, which was 200 g, and it was, therefore, necessary to approach it close enough to start moving. Due to the fact that it was a movement, it was necessary to carry out the measurement very carefully and precisely, which is also indicated by the value of the standard deviation.

Based on the findings from the results of the experimental tests given in Section 3 and the measurements described above, we can state that extreme temperatures in the standard working range of magnets do not affect the values of their intensity and field size. They only affect the switching capacity of the magnetic contacts. This effect is undoubtedly caused by the heating of a part of the magnetic contact such as screws for connecting cables or other components. These expand due to the high temperature, and therefore the connection/activation of the safety magnetic contacts takes a longer time or a greater distance. If we consider even more modern safety magnetic contacts in which more electronics are used, the distance could be even greater, because the temperature could affect the responsiveness of the safety magnetic contact, which would not be able to actively respond to closing or opening.

In the study, we also relied to a large extent on the results of the study by Calin and Helerea [14], but we were not able to confirm or refute their results, because our studies were significantly different. In our case, we focused on the switching ability of safety magnetic contacts, formed by two parts. On the contrary, the authors of the mentioned study measured magnetic quantities for one separate piece of the permanent magnet.

At first glance, it might seem that the magnetic contact, the USP 500, was also unreliable as it worked with a working distance almost 15 mm shorter. However, the opposite is true, as the magnetic contact worked correctly and within the operating range specified by the manufacturer. We have encountered a lack of working distances, namely that relatively wide ranges are mentioned and therefore the magnetic contact is considered reliable as long as it can be activated/switched on anywhere within the working distance.

The need to test the components of alarm systems is very important, as regular testing and evaluation of their reliability can predict the likelihood of their proper functioning [27,28]. In addition to the functional point of view, it is necessary to take into account the economic aspect, the value of which can be directly validated according to the needs of the secured object. Components of alarm systems are electrical devices for which the more expensive are not necessarily higher quality [29,30]. Knowledge of the probability of failure of a selected system component is very important for the effective design of the system. For example, if we know from the experimental tests the weaknesses of a selected component, we can increase its resistance by adding another component working on a different principle, thus achieving the component guarding the component. By partially increasing the security of a selected part of the building, we can increase the security of the entire building, according to its needs. The needs for object protection are an important input element in the evaluation and design of its security.

Knowing the reliability of individual system components is important because it allows us to choose from the same components on the market. All components should be certified and therefore usable in the conditions of the Slovak Republic. However, the certification takes place according to a predetermined methodological procedure, considers selected options for overcoming, and, last but not least, is performed under laboratory conditions. However, the potential perpetrator will not follow the methodological procedures or the technical standard permitted tools or objects when trying to overcome the security of the object [31,32]. Testing of alarm system components and detectors must be carried out as much as possible under non-standard conditions because only then is it possible to identify a higher degree of their reliability. Within the given partial part of the research of magnetic contacts, we focused on the simulation of extreme ambient conditions, as we encountered incorrect functionality of components several times. System malfunctions caused by extreme conditions can result in two basic possibilities. The first is the occurrence of false alarms and the second is the incorrect functioning of the system and thus the absence of guarding. In the first case, it is highly likely that the owner will be dissatisfied with the system due to multiple false alarms and could even partially deactivate the system, bringing us to the second option [33]. It is possible to test whether magnetic contacts or other components of alarm systems function according to the requirements of technical standards, but this would probably be unnecessary and we would get confirmation that they meet the requirements of the standard, as most components are certified.

5. Conclusions

Magnetic contacts are among the basic and most commonly used components of alarm systems. Their correct functionality is an integral part of the mantle protection of each secured object. As with other alarm system components, magnetic contacts are considered electrical devices and as such may not always achieve the required reliability. Reliability is often affected by restrictive conditions such as ambient temperature, poor installation, improper selection, and the like. As part of a comprehensive study of the reliability of magnetic contacts, we have decided to focus on the impact on the environment, as in recent years we have witnessed large fluctuations caused by global warming. Our goal was not to try to overcome safety magnetic contacts but to know their reliability at extreme temperatures, so we decided to perform only these types of measurements and avoided the use of parasitic magnets or other short-circuit options that can be used to overcome them.

We tested five magnetic contacts designed for external environments, environment classes III and IV, i.e., operating temperatures from −25 °C to +50/60 °C. In the tests themselves, however, we considered temperatures outside the temperature range. The temperature range we chose was from −40 °C to +70 °C within which we tested all components to achieve the same conditions. Through testing, we have found that temperature ranges in some cases cause a problem and, for example, in the case of magnetic contact USP

500, a shift of switching on at negative temperatures of 20 mm. In four of the five tested magnetic contacts, we observed a hysteresis phenomenon, which we could sometimes call directly proportional to the influence of the environment. We have found that magnetic contacts can work even in extreme conditions, but these to some extent affect their detection ability, which can decrease due to very low temperatures and increase at extremely high temperatures. Thus, in addition to the hysteresis, we can also observe the expandability of the material. Paradoxically, the magnetic contacts placed in the plastic housing showed a lower degree of hysteresis than the metal ones. To be able to carry out such tests, we created a simple test device, which, in addition to protecting the lives of researchers, also eliminated possible errors caused by heat radiation from the human body.

The magnetic contact behaved best in the tests with the type designation MAS 333, which during the change of temperatures showed an almost constant closing distance and the value of the hysteresis phenomenon was in the range of a maximum of 2 mm. On the contrary, the worst hit was the magnetic contact with the type designation SA 220, which in the case of switching on showed a difference of almost 30 mm and opening of almost 20 mm.

Through experimental tests, we probably came to the direct influence of the magnetic contact design on its reliability, as all square-shaped magnetic contacts showed good properties and activation within the working distances defined by the manufacturer or within a tolerance of 10%. Even the statistical deviation in these measurement cases was at a relatively low level, which can be considered as an effective measurement result. We recorded surprising results in the case of positive, high temperatures within which the magnetic contacts continued to work correctly and there was no rapid change in reliability or detection distance. Undoubtedly significant was the fact that we performed measurements in the temperature range up to 80 °C, i.e., the temperature considered critical. After this temperature, there could be a permanent change in the functionality of the magnetic contact or the permanent magnet. In the future, we would like to focus on this type of experimental test. The behavior of the hardened plastic into which the individual parts of the safety magnetic contacts are caught is also questionable.

It would also be possible to build on the results of our research and supplement the research with a comprehensive measurement of magnetic induction at high temperatures. This would confirm or refute the conclusions about the expansion of the materials used in the switching part of the magnetic contacts. The advantage, and at the same time disadvantage, is the fact that there are countless safety magnetic contacts on the market and, to fulfill our long-term goal, to amend the technical regulations. It will therefore be necessary to carry out a huge number of measurements to confirm the findings from camera systems and now also magnetic contacts, which speak to the fact that these devices can work fully beyond the range of operating temperatures according to the manufacturer.

Author Contributions: Conceptualization, M.B., A.V., V.S. and J.D.; methodology, M.B. and A.V.; software, V.S.; validation, M.B., A.V., V.S. and J.D.; formal analysis, M.B.; investigation, M.B., A.V., V.S. and J.D.; resources, V.S.; data curation, A.V. and J.D.; writing—original draft preparation, M.B.; writing—review and editing, A.V., V.S. and J.D.; visualization, M.B.; supervision, M.B.; project administration, M.B., A.V. and V.S.; funding acquisition, J.D. All authors have read and agreed to the published version of the manuscript.

Funding: This research was funded by Slovak research and development agency grant number APVV-17-0014 Smart tunnel.

Data Availability Statement: Data is contained within the article.

Conflicts of Interest: The authors declare no conflict of interest.

References

1. Explainthatstuff. Available online: https://www.explainthatstuff.com/howreedswitcheswork.html (accessed on 2 February 2021).
2. Mocholi Belenguer, F.; Martinez-Millana, A.; Mocholi Salcedo, A.; Arroyo Nunez, J.H. Vehicle Identification by Means of Radio-Frequency-Identification Cards and Magnetic Loops. *IEEE Trans. Intell. Transp. Syst.* **2020**, *21*, 5051–5059. [CrossRef]

3. Primin, M.A.; Nedayvoda, I.V. Non-Contact Analysis of Magnetic Fields of Biological Objects: Algorithms for Data Recording and Processing. *Cybern. Syst. Anal.* **2020**, *56*, 848–862. [CrossRef]
4. Ponce, E.A.; Leeb, S.B.; Lindahl, P.A. Know the Flow: Non-contact Magnetic Flow Rate Sensing for Water Meters. *IEEE Sens. J.* **2021**, *21*, 802–811. [CrossRef]
5. Lovecek, T.; Velas, A.; Durovec, M. *Bezpečnostné Systémy—Poplachové Systémy*; EDIS: Žilina, Slovakia, 2015; p. 230.
6. Velas, A.; Boros, M. *Intruder Alarms*; EDIS: Žilina, Slovakia, 2019; p. 113.
7. Parchomiuk, M.; Zymmer, K.; Domino, A. Security Systems of Power Converter Interface with Superconducting Magnetic Energy Storage for Electric Power Distribution. In Proceedings of the Conference on Progress in Applied Electrical Engineering, Koscielisko, Poland, 18–22 June 2018.
8. Kutaj, M.; Boros, M. Development of educational equipment and linking educational process with research. In Proceedings of the International Conference on Education and New Learning Technologies, Barcelona, Spain, 3–5 July 2017.
9. Kutaj, M.; Velas, A. Magnetické kontakty—Testovanie spol'ahlivosti. In *Riešenie Krízových Situácií v Špecifickom Prostredí*; EDIS: Žilina, Slovakia, 2015. Available online: https://new.fbi.uniza.sk/uploads/Dokumenty/weby/rks-archiv/2015/articles/Kutaj_Velas.pdf (accessed on 5 February 2021).
10. EN 50131-1. *Alarm Systems. Intrusion Systems. Part 1: System Requirements*; CENELEC: Brussels, Belgium, 2007. Czech equivalent available on the basis of a license for the Faculty of Security Engineering.
11. EN 50131-2-6. *Alarm Systems. Intrusion and Hol-up Systems. Part 2-6: Opening Contacts (Magnetic)*; CENELEC: Brussels, Belgium, 2009. Czech equivalent available on the basis of a license for the Faculty of Security Engineering.
12. EN 50130-5. *Alarm Systems. Part 5: Environmental Test Methods*; CENELEC: Brussels, Belgium, 2012; Czech equivalent available on the basis of a license for the Faculty of Security Engineering.
13. SHMU. Available online: http://www.shmu.sk/sk/?page=2049&id=928 (accessed on 5 February 2021).
14. Calin, M.D.; Helerea, E. Temperature influence on magnetic characteristics of NdFeB permanent magnets. In Proceedings of the International Symposium on Advanced Topics in Electrical Engineering, Bucharest, Romania, 12–14 May 2011.
15. Shahbazi, A.; Aghaei, J.; Pirouzi, S.; Shafie-khah, M.; Catala, J.P.S. Hybrid stochastic/robust optimization model for resilient architecture of distribution networks against extreme weather conditions. *Int. J. Electr. Power Energy Syst.* **2021**, *126*, 106576. [CrossRef]
16. Boroš, M.; Kučera, M.; Lenko, F.; Velas, A. Possibilities of practical tests of traffic video surveillance systems against external weather effects. In Proceedings of the International Scientific Conference on Transport Means, Palanga, Lithuania, 4 October 2019.
17. Boros, M.; Kucera, M.; Velas, A.; Valouch, J. Possibilities for experimental testing of alarm transmission systems. *Commun. Sci. Lett. Univ. Zilina* **2020**, *22*, 123–128.
18. Velas, A.; Lovecek, T.; Valouch, J.; Dworzecki, J.; Vnencakova, E. Testing radio signal range of selected components. *Commun. Sci. Lett. Univ. Zilina* **2018**, *20*, 68–77.
19. Penaska, M.; Velas, A. Possibilities of tracking city indicators in the sense of the Smart city concept. In Proceedings of the International Scientific Conference on Sustainable, Modern and Safe Transport, Zilina, Slovakia, 29–31 May 2019.
20. Boros, M.; Lenko, F. Possibility of transmission system disruption by intruder. In Proceedings of the International Scientific Conference on Sustainable, Modern and Safe Transport, Zilina, Slovakia, 29–31 May 2019.
21. Hasan, G. Towards an Ultra-Sensitive Temperature Sensor for Uncooled Infrared Sensing in CMOS–MEMS Technology. *Micromachines* **2019**, *10*, 108.
22. Equipment for Laboratories VTL and VCL for Temperature and Climatic Tests. Available online: https://www.labmakelaar.com/fjc_documents/7761_VIT_VTL%20VCL%20(E)%5b1%5d.pdf (accessed on 15 November 2020).
23. Hudec, I. Testovanie Vplyvu Prostredia na Detekčnú Schopnosť Magnetických Kontaktov. Diploma Thesis, University of Zilina, Zilina, Slovakia, 2019.
24. Boros, M.; Velas, A.; Lenko, F. Utilization of programmable microcontrollers to assess the reliability of alarm systems. In Proceedings of the International Conference on Diagnostics in Electrical Engineering, Pilsen, Czech Republic, 1–4 September 2020.
25. He, C.; Dou, W.Q.; Liu, X.C.; Yang, M.; Zhang, R.F. A Sensitivity Mapping Technique for Tensile Force and Case Depth Characterization Based on Magnetic Minor Hysteresis Loops. *Chin. J. Mech. Eng.* **2020**, *33*, 84. [CrossRef]
26. Tomioka, K.; Uchida, K.I.; Iguchi, R.; Nagano, H. Non-contact imaging detection of thermal Hall effect signature by periodic heating method using lock-in thermography. *J. Appl. Phys.* **2020**, *128*, 215103. [CrossRef]
27. Mishra, B.; Smirnova, I. Optimal configuration of intrusion detection systems. *Inf. Technol. Manag.* **2021**, 1–14. [CrossRef]
28. Lovecek, T.; Velas, A.; Kampova, K.; Maris, L.; Mozer, V. Cumulative Probability of Detecting an Intruder by Alarm Systems. In Proceedings of the International Carnahan Conference on Security Technology Proceedings, Medellin, Colombia, 8–11 October 2013.
29. Kampova, K.; Makka, K.; Zvarikova, K. Cost benefit analysis within organization security management. In Proceedings of the International Scientific Conference Globalization and Its Socio-Economic Consequences—Sustainability in the Global-Knowledge Economy, Rajecke Teplice, Slovakia, 9–10 October 2020.
30. Soltes, V.; Stofkova, K.R.; Lenko, F. Socio-economic consequences of globalization on the economic development of regions in the context of security. In Proceedings of the International Scientific Conference Globalization and Its Socio-Economic Consequences—Sustainability in the Global-Knowledge Economy, Rajecke Teplice, Slovakia, 9–10 October 2020.

31. Liu, H.; Yue, X.H.; Ding, H.; Leong, G.K. Optimal Remanufacturing Certification Contracts in the Electrical and Electronic Industry. *Sustainability* **2017**, *9*, 516. [CrossRef]
32. Zhang, R.; Liu, M.; Wang, X.; Gao, E.; Lü, J.; Wang, C.; Kan, F.; Wang, B. Research on Electrical Technology Course Reform based on Professional Certification of Engineering Education. In Proceedings of the International Conference on Education, Management, Computer and Society, Shenyang, China, 1–3 January 2016.
33. Roohi, M.H.; Chen, T.W. Generalized moving variance filters for industrial alarm systems. *J. Process Control* **2020**, *95*, 75–85. [CrossRef]

micromachines

MDPI

Article

New Possibilities for Testing the Service Life of Magnetic Contacts

Martin Boroš *, Andrej Vel'as, Zuzana Zvaková and Viktor Šoltés

Department of Security Management, Faculty of Security Engineering, University of Zilina, Univerzitna 8215/1, 010-26 Zilina, Slovakia; andrej.velas@fbi.uniza.sk (A.V.); zuzana.zvakova@fbi.uniza.sk (Z.Z.); viktor.soltes@fbi.uniza.sk (V.Š.)

* Correspondence: martin.boros@fbi.uniza.sk; Tel.: +421-41-513-6668

Abstract: Magnetic contacts we could define as a switching device used in transport structures such as a tunnel, to which the manufacturer prescribes a certain number of closures within its lifetime, during which they should operate flawlessly. Verification of the data provided by the manufacturer is time-consuming and physically demanding due to the data being large in number. For this reason, we developed a test device using torque in the research of magnetic contacts, which greatly automates the whole process and thus eliminates human error. The test device can use internal memory to calculate the number of closures of magnetic contacts and then transmit the digitized data. The test device is registered as an industrial utility model and can be used to test any magnetic contacts.

Keywords: magnetic contacts; torque; the calculation in memory; automation

Citation: Boroš, M.; Vel'as, A.; Zvaková, Z.; Šoltés, V. New Possibilities for Testing the Service Life of Magnetic Contacts. *Micromachines* **2021**, *12*, 479. https://doi.org/10.3390/mi12050479

Academic Editor: Viktor Sverdlov

Received: 26 February 2021
Accepted: 19 April 2021
Published: 22 April 2021

1. Introduction

Reliability is a natural property of products to perform the activities for which they were created, under predefined conditions and for a defined period. Reliability can be assessed for almost all devices in different types of industries. The parameter determining the service life of the equipment is also connected with reliability. We can divide the service life into two groups. One group consists of devices that define the reliability of time, i.e., the time during which the manufacturer declares their correct functionality, such as LED bulbs for which the manufacturer declares proper functionality somewhere at 50,000 h of use for lighting. The second group consists of switching devices in which the service life is defined in terms of the maximum number of switchings, such as conventional light bulbs for which the manufacturers declare the service life somewhere at the level of 10,000 switchings. In practice, this means that in the case of the first option, the LED bulb should stop shining after the time specified by the manufacturer, and in the case of the second option, if we turn the bulb off and on in half the number specified by the manufacturer, it should also stop shining or lighting up [1–4]. The service life of electrical equipment, in particular, has been a very inflected concept in recent years, as some are convinced that manufacturers purposefully install a time-memory component in equipment that is activated after a certain period of use, usually after the warranty period.

An exception from the point of view of reliability is also the electrical devices or components forming an electrical security system designed to identify and detect the presence or attempt of an intruder to enter the building. Even in this case, we could apply the above distribution of life because while the passive infrared detector must be able to detect the guarded area for a long time, the magnetic contact used to secure windows and doors has a predefined value of the number of closures from the manufacturer [5–8].

Magnetic contacts are used in almost every industry for various purposes, for example, to detect the opening of switch cabinets and switchboards in electronics, as an opening mechanism in keyless furniture designs, in the toy industry, and many others. From the point of view of security, as mentioned, they have their fixed place in the electrical security system where they are used as opening detectors, i.e., they sound an alarm in case of unauthorized opening of windows, doors, gates, or blinds or shutters on windows. Depending on the nature and requirements of the secured object, it is possible to create complete protection from magnetic contacts by fitting them on all doors and windows. The magnetic contacts are formed by a permanent magnet and a contact, also called a reed contact. The reed contact consists of a sealed glass tube filled with a protective atmosphere in which two ferromagnetic contacts are placed (Figure 1). The magnet is installed on a moving part of a window or door and a reed contact on a fixed part, i.e., a door frame or frame. At the moment when the permanent magnet is close to the tongue contact, it is in the basic state, usually closed, due to the magnetic field of the permanent magnet. When the magnet is moved away, for example, due to the opening of a door or window, the magnetic field affecting the tongue contact weakens, and the contact changes its state to the opposite. If the contact is of the normally closed type, it opens, and if it is of the normally open type, it closes [7,9–12].

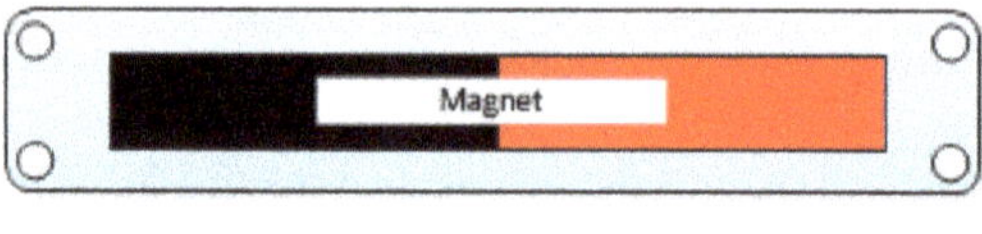

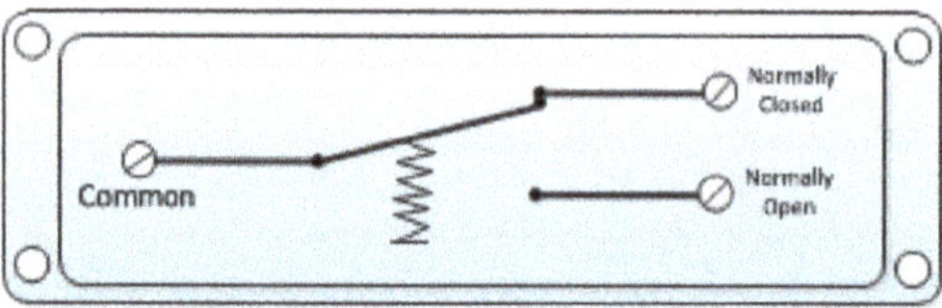

Figure 1. Principle of operation of magnetic contact [7].

The reliability or correct functionality of the magnetic contacts can be tested according to predefined possibilities, as specified in more detail in the European technical standard with the type designation EN 50131-2-6 [13]. The technical standard prescribes several possibilities, or types of tests where the main one is the resistance of the magnetic field to the external magnetic field generated by the parasitic magnet. In this case, it is determined whether the electric security system correctly evaluates the attempt to disrupt or sabotage the system. In addition to the parasitic magnet test, the technical standard defines other tests, such as the basic detection function test in which the values of closing and opening of the magnetic contact are verified and then compared with those specified by the manufacturer. However, these and other types of tests have one basic disadvantage/disadvantage and therefore, given the possibilities of today's market, they are relatively strict and do not take into account the possibilities of various commonly available magnets [10,13]. The second mentioned test, or the basic detection function, could be described as a direct verification of the reliability of the magnetic contact. This is because it takes place in a slow, gradual, and simulated movement from the closed position of the magnetic contact, with the magnet facing away from each other, thus determining the actual value of the opening. Subsequently, the whole process is repeated, but in the opposite direction, which determines the value of closing the magnetic contact. The same procedure can be performed in the case of the oblique direction of the permanent magnet, as it should be able to work in all directions of movement. These types of tests are relatively demanding and in many cases even lengthy, which increases the risk of error of the human factor, i.e., the person performing the tests themselves. One of the basic and challenging tasks is to accurately determine the value of individual distances. To avoid errors and inaccuracies in the test, it is recommended that graph paper be used to determine the values better and faster. To know when the magnetic contact closes so that it is activated, it is necessary to connect a signaling device to it. In the absence of a signaling device, it is possible to use a complete installation of an electrical security system with a control panel, with which we can know when the contact was activated [14–17].

Within the professional community, it is possible to observe several research activities aimed at measuring the reliability of magnetic contacts, but in all cases they represent only a partial part of the research. For example, Gong and his team focused on measuring the sorting of magnets in factory production, in which they pointed to the lack of homogeneity of the magnetic field. In one of their studies, the Carus team focused on improving the efficiency of electric drives using permanent magnets, within which the goal was to define a uniform procedure for measuring the impact on control algorithms [11,18,19].

Despite measuring and testing the reliability of magnetic contacts, it is always a question of using magnetic contacts, as mentioned, and not the switching capacity itself. Therefore we decided to research the total magnetic contacts for security needs to pay attention to their actual switching functions, and within it and its recurrence.

As mentioned above, magnetic contacts are one of the basic components of electrical security systems designed to primarily protect the building envelope. To indicate the danger, they use the influence of magnetic induction, by which the permanent part is attracted to the reed contact, which is currently supplemented by other electronics, such as an LED designed to indicate the closing/opening of the magnetic contact. The reliability of magnetic contact from a safety point of view is understood as to whether the opening and the opening of the window or door leaf caused by it is detected, as the magnetic contact needs to be supplemented with components in other layers of protection, such as spatial or object protection, for the complex security of the building. It is also highly important to use mechanical means of restraint to make it more difficult for the intruder to enter and escape from the secured object. It is necessary to realize that even the highest quality component of the electrical security system can be surpassed, and this depends only on the equipment and skills of the intruder. Therefore, in the professional and public sphere, we more often encounter the term probability, and if we consider a comprehensive assessment of the secured object, we use the term cumulative probability of intruder detection [20]. The cumulative probability of intruder detection is calculated using Formula (1) [6].

$$Pcp = [1 - \prod_{i=1}^{n} (1 - Pd)] * Pats * Pf * Phf \quad (1)$$

where *Pcp* is the cumulative probability of intruder detection, *n* is the number of detection zones of the object, *Pd* is the probability of detection by active elements (magnetic contacts, passive infrared detectors or PID), *Pats* is the probability of transmitting intrusion information to the monitoring center, *Pf* is the probability of a fault-free state of the system, and *Phf* is the probability of human factor failure (guard, technician, system administrator).

To express the correct value of the cumulative probability, it is necessary to know as many input parameters as possible. In this case, it is the reliability of individual active elements installed in the building, i.e., detectors, which are designed to detect the intruder. As this is a large-scale issue, it is necessary to address it in the long term and extensively. In terms of time, such large-scale measurements could take several years if carried out by a single research team. Therefore, it is appropriate to divide the research activities related to the experimental measurements between several research teams that can work in parallel. As part of testing the reliability of individual components, data for PID [21] and transmission systems [22] were created based on experimental tests. PID detectors were chosen because of their frequent installation, as they are very often installed and they generate an alarm when detecting the movement of the intruder, and as a secondary reason because they are primarily used in spatial protection. We therefore chose the procedure in terms of selecting one main representative of the zone, for which we express the values of reliability. Another very important component is the magnetic contacts, which need to be tested as much as possible.

2. Materials and Methods

The overall research of the reliability of magnetic contacts from the point of view of safety could be divided into two groups of measurements. The first focuses on the accuracy of detection, activation, and deactivation of magnetic contact, and the second focuses on the life of magnetic contacts, i.e., the frequency of their connection or disconnection. In the case of the first group of measurements, we determined the exact procedure that we followed for all measurements as well as their repetitions. The basis was the division of this part into a part of closing, i.e., activation, and the time of expansion, i.e., deactivation of the magnetic contact. The essence of the measurement was the smooth movement of the magnet in the direction and against the axis of the magnetic field to express the approach and departure distance. After reaching the approach distance, a rest signal must be generated within the electrical security system, and after reaching the distance, an alarm signal must be generated [13,23]. The mentioned measurement procedure consisted of the following parts:

1. Becoming familiar with the magnetic contacts by studying the package leaflet from the manufacturer;
2. Performing the recovery and subsequent connection of magnetic contacts as well as other measuring components;
3. Realizing the measurements of the closing mode and opening mode;
4. Repeating each measurement 30 times for version 1 and 1000 times for version 2;
5. Calculating the arithmetic mean (x) of closing and opening of the magnetic contact using the relation $\mathrm{x} = \frac{\sum_{i=1}^{n} x_i}{n}$, then rounding the values to whole numbers or halves upwards;
6. Comparing the results obtained with the data provided by the manufacturers.

The second group of measurements had essentially a very similar procedure, but the difference occurred in the case of point number 5, as the arithmetic mean was not calculated but only the total closing/opening of the magnetic contact was calculated. The aim of this type of measurement was not to know the working distance of the magnetic contact but its reliability at a preset distance. In the case of these measurements, we relied on a technical standard that allows the test requirements to be met with a tolerance of +/−10% [13].

To facilitate the measurement process, we made a test device or test connection, which had the task of indicating the closing/opening of the magnetic contact. During the measurement, we even created multiple connections for a more comfortable and accurate result. Version 1 of test circuit 1 consisted of the use of a multimeter marked UNI-T UT70A, in which we used the possibility of circuit integrity, the so-called short-circuit probe, a ruler, and graph paper. The circuit diagram of test circuit 1 is shown in Figure 2.

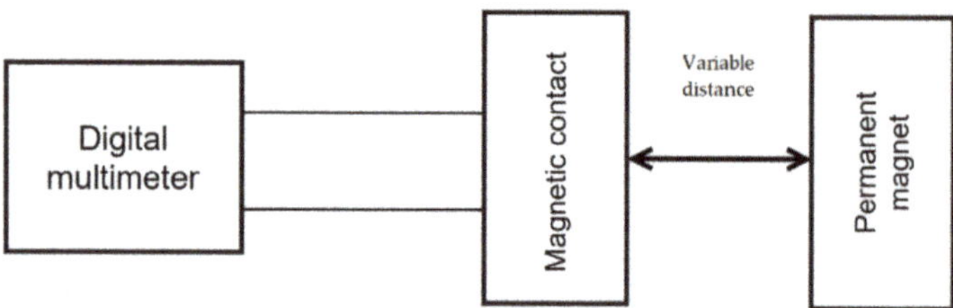

Figure 2. Circuit diagram of the testing version 1.

After the implementation of the pilot 30 tests, we proceeded to the use of another measurement option, in which we decided to engage another sensation to identify the activation of the magnetic contact, namely sight. We therefore created a connection with an LED intended for visual inspection of the change in the state of the magnetic contact (version 2).

Version 2 consisted of the electrical circuit, as shown in Figure 3, consisting of a laboratory power supply, a Yihua HY1503D, a 560 Ω resistor, and a color LED (red in our case). The output conductors, which were connected directly to the magnetic contact, switched the negative branch of the electrical circuit [23,24].

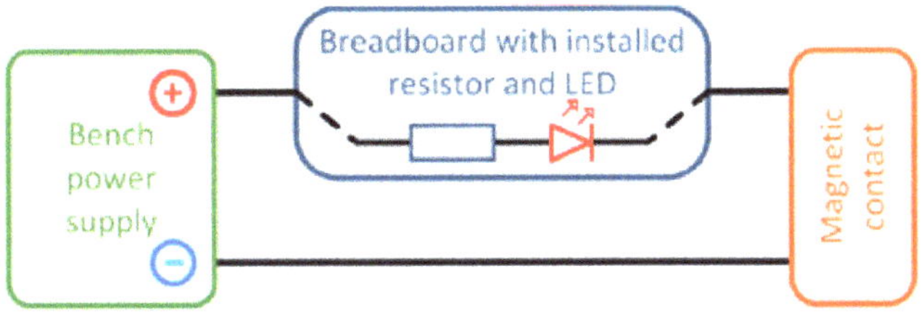

Figure 3. Circuit diagram of the testing version 2 [10].

With the newly created test facility, we also conducted 30 pilot tests to determine which test facility is more accurate. However, the connection created was relatively impractical, mainly due to the large size of the laboratory source. Apart from the impracticality, from the source's point of view, the measurement was done in an uncomfortable manner because we held both parts of the magnetic contact in our hands. We therefore decided to create an upgrade of the test equipment, which consisted of a fixed and a sliding part into which the individual parts of the magnetic contacts were installed. We made these parts from polymethyl methacrylate parts. Specifically, it was an "L" shaped profile in which a magnetic contact was installed. In addition to this, a contact field with a resistor and an LED was placed on the profile to indicate switching on and the power supply, which we solved using a backup power supply, consisting of two alkaline batteries marked LR03 with a nominal voltage of 1.5 V. A permanent magnet was attached to the second part, which is a moving part. This part was fully mobile and only added to the L profile. The advantage of this type was that the reading of the distance was instantaneous since on the L profile there was a scale of the ruler, the beginning of which is point 0, which was situated at the end of the fixed part and thus also of the magnetic contact. The circuit diagram of the improved version of the test circuit version 3; without the polymethyl methacrylate, the construction is shown in Figure 4 [13,23,25].

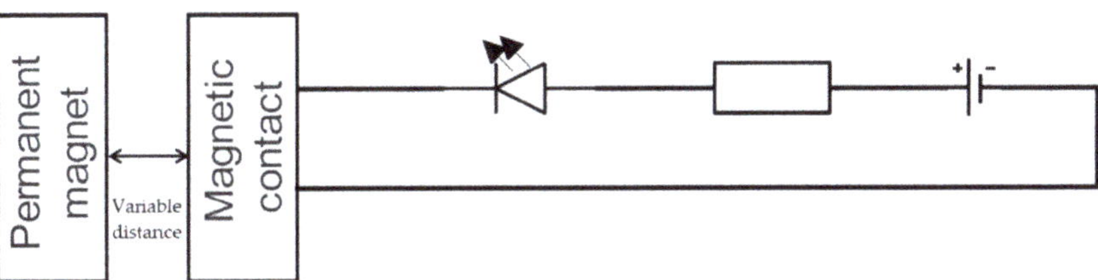

Figure 4. Circuit diagram of the testing version 3.

In the case of the second group of measurements, i.e., the lifetime of magnetic contacts, we devised a fully functional device that is protected by the Industrial Property Office as an industrial utility model. The essence of the test device is the possibility of performing a continuous rotary or oscillating movement, the basis of which is a twisting moment formed by a low-speed, motor drive. This was located in the lower part of the test device, together with the other technological part intended to express the number of repetitions, time recording, and the possibility of setting the speed, i.e., the speed of movement. With the help of the test device, it is possible to fully automate the measurement at a predetermined distance. This test device also consisted of two parts: the lower part with an electric motor, and the technical part was a fixed part to which the magnetic contact was attached. The movable part was formed by a plate that rotated around its center in the direction and speed asset. A permanent magnet was installed on its edge at the working distance of the magnetic contact. Ideally, the working distance was set approximately in the middle of the scale specified by the manufacturer or if the actual working distance from previous tests is known. Subsequently, the measurement was performed by successive rotations of the plate, and each time one circuit was performed, the given value was recorded in the internal memory of the device; also in a given number of rotations, the time is determined and whether the magnetic contact has been activated. The recorded data can then be downloaded from the test device to a portable device in the format of a file with the final .xlsx file, which can be edited in MS Excel and the measurement of which can be evaluated. In Figure 5, we can see the test device with the magnetic contact placed before starting the measurement [23,26].

Figure 5. Magnetic contact life tester.

The starting position for testing magnetic contacts was formed by pilot measurements, in which in addition to the measured values, we also evaluated the reliability of the given trial versions. Four commonly used magnetic contacts with type designations were selected for mutual comparison: Bestkey BP-1013, Bestkey BS-2013, SUNWAVE SD 8561, and USP 130SP.

These selected magnetic contacts were sequentially tested in all three versions of the measurement sets described in more detail. We decided to perform an extensive complex measurement focused on both the reliability and durability of magnetic contacts. In the initial first phase, we implemented a pilot of 30 measurements through all three test connections to determine the most accurate one. Subsequently, in the second phase, we performed 1000 repetitions with each magnetic contact using the most effective test circuit. In the final and third phase, we performed the lifetime measurement for 10,000 magnetic trials, for one magnetic contact, using the last described test device. For each measurement, the arithmetic mean of the measured values was determined, rounded up to whole numbers or halves. We obtained these values using the average function in MS Excel. We also determined the standard deviation for each measurement using the STDEV function in MS Excel. The overall results are shown in the following section.

3. Results

The first phase, as mentioned, was to test all the proposed test connections. The measurement was performed in two directions, first from point 0, a connected magnetic contact and a permanent magnet, and the subsequent removal of these parts, until the LED goes out and the distance is recorded. Subsequently, we moved the parts of the magnetic contact to a distance of 10 cm and gradually had them approach each other until the LED light came on. The values measured by the individual methods for the magnetic contacts tested are shown in Table 1 together with the standard deviation.

Table 1. Results of measurements with a multimeter.

Type of Magnetic Contact	Datasheet Distance [mm]	Test Connection (Version 1)				Test Connection (Version 2)				Test Connection (Version 3)			
		Clamping Distance [mm]	Standard Deviation	Opening Distance [mm]	Standard Deviation	Clamping Distance [mm]	Standard Deviation	Opening Distance [mm]	Standard Deviation	Clamping Distance [mm]	Standard Deviation	Opening Distance [mm]	Standard Deviation
Bestkey BR1013	0–25	35	3.18	51	4.14	29.5	2.74	38.3	2.54	24	2.67	31	1.59
Bestkey BS2013	0–31	39	3.38	46	2.83	28	4.68	38	2.95	15.5	3.69	22.6	1.31
SUNWAVE SD 8561	12.7–25.4	24	4.28	46.5	4.05	28.5	4.81	38	4.48	28.1	1.67	36.1	2.53
USP 130SP	0–25	30.5	2.71	46.5	3.99	30	4.76	30	2.05	27.5	3.89	29	1.83

From the measured results, it is clear that the magnetic contacts do not correspond to the data given by the manufacturer in the technical documentation. In addition, in all four cases, we can observe the fact that the value of clamping itself is higher than the values given by the manufacturer. The opening values, which are a more important indicator than the closing ones, are in some cases even twice as high as the values given by the manufacturer [27].

Even the results in version 2 are not paradoxically better than those with version 1, but there is a slight shift to the values given by the manufacturer. On the positive side, the values of closing and opening decreased significantly by an order of 15 mm. The deviation can also be caused by the error rate of the human factor and the complexity of gripping the permanent magnet during measurement.

The final, third type of test device was a product in which the permanent magnet could be attached to the sliding part, thus increasing the comfort of measurement.

As we can see, the measured values using the test circuit version 3 are the closest to those specified by the manufacturer in the datasheet. We managed to measure, in the case of the magnetic contact Bestkey BS2013, the values corresponding to those from the datasheet. In some cases, for the Bestkey BR1013 we managed to reduce the value by 7 mm in case of opening, and in case of switching on we reached the level of the manufacturer. In other cases, the worst effect was the Sunwave magnetic contact, in which case we are far behind the values given by the manufacturer in this measurement.

A separate group of results is the values of the standard deviation, which in some cases are relatively high and almost at level 5. The smallest value is reached by the standard deviation in the case of the third version, so based on these data we can say that this version is the most reliable.

After completing the first phase followed by the second, we focused on 1000 repetitions using test circuit version 3. Gradually, we implemented all the repetitions for both the closing and opening of the magnetic contacts. Due to the complexity of the measurement from the point of view of attention, we took a break that lasted at least half an hour. The results, obtained together with the expression of the standard deviation, are shown in Table 2.

Table 2. Measured values at 18,000 repetitions.

Type of Magnetic Contact	Datasheet Distance (mm)	Clamping Distance (mm)	Standard Deviation	Opening Distance (mm)	Standard Deviation
Bestkey BR1013	0–25	23.5	1.99	25.5	2.01
Bestkey BS2013	0–31	27	1.72	29.5	1.86
SUNWAVE SD 8561	12.7–25.4	15.5	2.06	23.5	2.54
USP 130SP	0–25	22.5	1.47	25	1.69

As we can see from the results given in Table 2, we were advised to achieve the average values, which fall within the intervals given by the manufacturer, using the test circuit version 3. The only discrepancy occurred in the case of the opening distance for the magnetic contact Bestkey BR1013, for which the distance is exceeded by 0.5 mm, i.e., a negligible distance. The values of the standard deviation are, except for one magnetic contact, at a very good level. According to the results, the problematic or faulty magnetic contact is SUNWAVE SD 8561. This magnetic contact has shown an appropriate degree of error or unreliability since the beginning of testing. However, it should be noted that despite the high degree of error, we measured either below or within the range specified by the manufacturer.

The last phase of testing was focused on experimental tests using an autonomous testing device, i.e., the complete construction of which is protected by the Industrial Property Office of the Slovak Republic. The device, as mentioned, uses the torque from the electric motor, the power of which and thus the torque can be adjusted using a potentiometer. During testing, the same conditions were ensured for each repetition, i.e., the distance as well as the speed of movement of the revolution was always identical for all repetitions. In our case, one turn lasted half a second, during which a step change occurred due to the closing of the magnetic contact. The recording was performed twice. In the first case it was a matter of counting one rotation of the rotating part about its axis, and in the second case, we counted whether the closing of the magnetic contact occurred. Subsequently, the data are recorded in the MS Excel software program, and it was therefore possible to clearly define in which rotation occurred and in which the closing of the magnetic contact did not occur. In the case of switching on, the value 1 was recorded, and if no switching was made, the value 0 was recorded. From these values, we plotted the switching on for magnetic contact USP 130SP shown in Figure 6. We gradually measured all magnetic contacts measured in the usual way in the previous expression. The measured values are given in Table 3. It should be noted that in these measurements, we used new pieces of magnetic contacts to avoid possible data distortion due to wear from previous measurements.

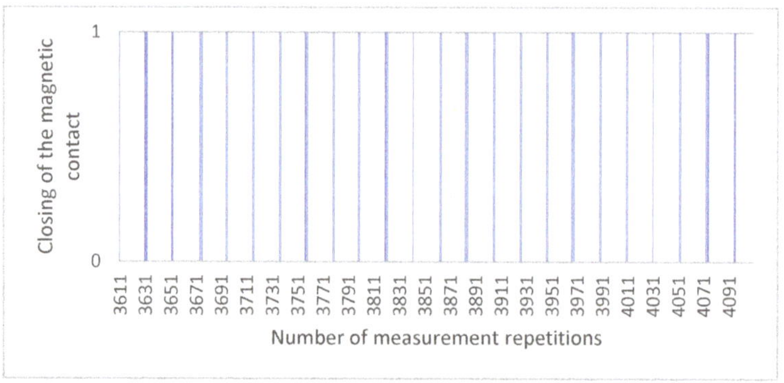

Figure 6. The data output of 500 repetitions from the test device (USP 130SP).

Table 3. Measured values using an autonomous test device.

Type of Magnetic Contact	Datasheet Distance [mm]	Set Distance [mm]	Number of Measurements	Activation of Magnetic Contact	Nonactivation of Magnetic Contact
Bestkey BR1013	0–25	22.5	10,000	9001	999
Bestkey BS2013	0–31	28	10,000	9104	896
SUNWAVE SD 8561	12.7–25.4	17	10,000	8913	1087
USP 130SP	0–25	22.5	10,000	9031	969

In the case of the last third phase, we decided to use the least erroneous USP 130SP. We subjected it to a load of similar 10,000 clamps at a predetermined working distance. For our measurement, we used the working distance achieved in the second phase, i.e., 22.5 mm. We measured the last phase on a test device, the export of which we subsequently evaluated in the MS Excel program. From the measured values we can state that the magnetic contact active in 9031 cases and in 969 not, which is within a tolerance of 10%, allowing within the technical standard. As the table with the data export would be confusing due to the size, we present the results in graphical form on Figure 6 for 500 randomly selected repetitions.

In Figure 6, we can see the step changes indicating the activation of the magnetic contact, where the initial position is 0 and the value 1 means the nonactivation of the magnetic contact in a given turn. The values in the figure do not reflect the duration of activation/non-activation of the magnetic contact; it only reflects the fact that the change has occurred. On the graph, 1 is specifically broken into 448 activations with a value of 0 and 52 nonactivations with a value of 1. The total measurement was 10,000 repetitions, and they lasted approximately two and a half hours.

From the results shown in Table 3, the deviations between the magnetic contacts are not at such a large level. Each of them was able to switch on (activate) approximately 9000 times, which represents a reliability of 90%. It should be noted that the activations represented a step change, a step duration of the magnetic pulse, for about half a second. With each rotation, the permanent magnet performed the same path and was always close to the reed contact for the same amount of time. The only difference in values could occur when activating the device.

4. Discussion

Pilot testing revealed several shortcomings, which we continuously eliminated, and with the help of the test connection version 3, we managed to realize 1000 repetitions of closing and opening of each tested magnetic contact. Together, in all phases of measurement, we performed 48,360 repetitions with two sets of four tested magnetic contacts. As mentioned, the most suitable connection was version 3, in which we did not have to hold parts of the magnetic contacts in our hands; they were placed on a test device, and thus this eliminated the possible error rate of an incorrect reading of the detection distance. The implementation of the second phase took us almost three days with breaks, along with the fact that we only measured during the day. The given figure is diametrically different from the value of two and a half hours, which was how long the third phase lasted.

All phases followed each other and played an important role; if we did not identify in the first phase the most suitable way to perform the measurement, we would probably obtain skewed results and incorrectly estimate the value of the working distance needed for phase 3. Of course, it is possible to enter the working distance as we saw in the results in Table 2, the actual working distances are different than stated by the manufacturer. It is appropriate to consider whether in the case of setting the working distance to 25 mm in phase 3, we would obtain a significantly lower value of not activating the magnetic contact. This reasoning is appropriate, as the mentioned value is marginal according to the manufacturer.

In the third phase of testing, we completely automated the measurement of the reliability of magnetic contacts using a test device. To further define and understand the problem, we implemented 3 phases, as the results from the second and third phases are comparable. The difference occurs in the implementation of tests in terms of the magnetic field of the permanent magnet. In the case of manual tests, we moved the permanent magnet in a vertical position away from and to the tongue contact, while monitoring the value of the distance of its closing. In the second case of measurements, the movement of the permanent magnet was performed in the base at a horizontal level as it was mounted on a movable part that rotated about its axis. During all tests, the same conditions were maintained in terms of temperature and humidity in the room, and so the values of the magnetic field strength were the same. The difference between the measurements can be seen in several factors. In the case of manual measurement, we had to be maximally focused, and we needed to focus on the accuracy of the reading value; in the case of autonomous equipment, we eliminated these conditions, which reduced the measurement error as we assumed that within the working distance specified by the manufacturer, the magnetic contact should be able to work fully. From the point of view of the reliability of magnetic contacts, we can conclude that with the help of an automated test device, the value of the reliability of magnetic contacts reached 90% or 0.9 in the case of dimensionless expression [5,28,29].

The measurement of the detection characteristic, i.e., the closing and opening of the magnetic contacts, is an important and basic prerequisite for expressing the probability of the detector being overcome by the intruder or overcoming the secured zone. If the intruder wants to achieve the identification of the protected interest, it is necessary to cross all detection zones of the protected object. To express the overall probability of intruder detection, a cumulative probability is used that takes into account the number of zones in the building, the probability of correct detection by electrical security system components, the probability of system failure, the probability of correct information transmission, and the probability of human failure. The total value of the cumulative probability should be close to 1. Knowledge of the total value of the probability is also necessary due to the design of the system from an economic point of view [6,13,30,31].

The magnetic contacts, the measurement of which is devoted to the article, belong to the category of the probability of correct detection by the components of the electrical security system. We performed experimental measurements mainly due to a more accurate expression of the probability, as its decrease depends on several factors, and one of them is the reliability of the component. To subsequently express the total probability of the detection zone, the reliability of all components installed in the zone is calculated. Paradoxically, however, it can be stated that to express the total value of the probability of crossing the detection zone, it is sufficient to express the reliability of one component. In reality, however, the use of one component in a given zone is insufficient and their combination is recommended [7,20,32,33].

5. Conclusions

The article is devoted to a comprehensive, experimental testing of safety magnetic contacts from the point of view of the correct functionality and service life of a component intended for the mantle protection of objects. Magnetic contacts are one of the cheapest and most commonly used components of an electrical security system. The measurements are part of extensive scientific research activity to express the cumulative probability of detecting an intruder of a protected object. By their nature, they follow the reliability of passive infrared detectors, transmission systems, and other parts of the electrical security system.

We performed the experimental measurement in three phases. In the first we created three ways of recording the real working distance of magnetic contacts. In the second phase, we performed an extensive measurement for each magnetic contact using the most efficient method from the first phase. The final, third phase was focused on a large number of switching repetitions for one magnetic contact. As part of the results of the first phase, we found that the most effective way to record the working distance of magnetic contacts is a test circuit version 3, which consisted of a backup power supply, LED, resistor, and a structure made of nonconductive material. Using this method, we performed a total of 8000 repetitions in the second phase. With a higher number of repetitions, we were able to measure the values of the working distance almost identical to those stated by the manufacturer in the technical documentation, and therefore we can state that the magnetic contacts are reliable.

In the third phase, a test device was used which eliminated the error rate caused by the human factor. We could therefore clearly define the percentage of which the tested magnetic contacts are reliable because if the closing did not occur under the same conditions, it meant the failure of the magnetic contact. Such a failure means a malfunction of the security system, i.e., insufficient protection of the object, which may be endangered by the intruder. In addition to clearly defining the reliability of magnetic contacts, we improved the ability to obtain data by automating the testing process, as the testing itself does not require supervision and can be performed at any time and to any extent. Subsequently, we can better know the cumulative probability of intruder detection in a charming object as well as knowledge of the theoretical basis for simulation programs designed to find the most effective route of an intruder.

Experimental testing of magnetic contacts is complemented by long-term research into the reliability of components of the eclectic security system, and it is necessary to repeat it regularly. The disadvantage of measuring the reliability of these types of components is the time-consuming nature of the whole process and the possible error rate of the human factor when reading the value of closing or opening of the magnetic contact. However, these parameters can be largely eliminated utilizing a test device designed by us to test the lifetime of magnetic contacts. Although this device has several shortcomings, these can be eliminated, improved, and can become even more efficient in measuring the reliability of magnetic contacts in the future.

Author Contributions: Conceptualization, M.B., A.V., Z.Z., and V.Š.; methodology, M.B. and A.V.; software, A.V.; validation, M.B., Z.Z., V.Š., and A.V.; formal analysis, M.B.; investigation, M.B., A.V., V.Š., and Z.Z.; resources, A.V.; data curation, V.Š. and Z.Z.; writing—original draft preparation, M.B.; writing—review and editing, A.V., V.Š., and Z.Z.; visualization, M.B.; supervision, M.B.; project administration, M.B., A.V., and V.Š.; funding acquisition, A.V. All authors have read and agreed to the published version of the manuscript.

Funding: This research was funded by Slovak research and development agency grant number APVV-17-0014 Smart tunnel.

Data Availability Statement: Data is contained within the article.

Conflicts of Interest: The authors declare no conflict of interest.

References

1. Vincent, M.; Chiesi, L.; Fourrier, J.C.; Garnier, A.; Grappe, B.; Lapiere, C.; Coutier, C.; Samperio, A.; Paineau, S.; Houzee, F.; et al. Electrical contact reliability in a magnetic MEMS switch. Electrical contacts 2008. In Proceedings of the Fifty-Fourth IEEE Holm Conference on Electrical Contacts, Orlando, FL, USA, 27–29 October 2008.
2. Burkov, A.F. Main Provisions of Theory of Reliability in Relation to Shipboard Electrical Equipment. In Proceedings of the International Conference on Industrial Engineering, Applications and Manufacturing (ICIEAM), Vladivostok, Russia, 16–19 May 2017.
3. Molina-Lopez, F.; de Araujo, R.E.; Jarrier, M.; Courbat, J.; Briand, D.; de Rooij, N.F. Study of bending reliability and electrical properties of platinum lines on flexible polyimide substrates. *Microelectron. Reliab.* **2014**, *54*, 2542–2549. [CrossRef]
4. Kiele, P.; Hergesell, H.; Bühler, M.; Boretius, T.; Suaning, G.; Stieglitz, T. Reliability of Neural Implants—Effective Method for Cleaning and Surface Preparation of Ceramics. *Micromachines* **2021**, *12*, 209. [CrossRef] [PubMed]
5. Kampova, K.; Lovecek, T.; Rehak, D. Quantitative approach to physical protection systems assessment of critical infrastructure elements: Use case in the Slovak Republic. *Int. J. Crit. Infrastruct. Prot.* **2020**, *30*, 100376. [CrossRef]
6. Lovecek, T.; Velas, A.; Kampova, K.; Maris, L.; Mozer, V. Cumulative Probability of Detecting an Intruder by Alarm Systems. In Proceedings of the International Carnahan Conference on Security Technology Proceedings, Medellin, Colombia, 8–11 October 2013.
7. Velas, A.; Boros, M. *Intruder Alarms*; EDIS: Slovakia, Žilina, 2019; p. 113.
8. Wang, M.; Shi, H.; Zhang, H.; Huo, D.; Xie, Y.; Su, J. Improving the Detection Ability of Inductive Micro-Sensor for Non-Ferromagnetic Wear Debris. *Micromachines* **2020**, *11*, 1108. [CrossRef] [PubMed]
9. Kim, C.H.; Kim, J.; Park, J.O.; Choi, E.; Kim, C.-S. Localization and Actuation for MNPs Based on Magnetic Field-Free Point: Feasibility of Movable Electromagnetic Actuations. *Micromachines* **2020**, *11*, 1020. [CrossRef] [PubMed]
10. Kutaj, M.; Boros, M. Development of educational equipment and linking educational process with research. In Proceedings of the International Conference on Education and New Learning Technologies, Barcelona, Spain, 3–5 July 2017.
11. Jia, S.; Peng, J.; Bian, J.; Zhang, S.; Xu, S.; Zhang, B. Design and Fabrication of a MEMS Electromagnetic Swing-Type Actuator for Optical Switch. *Micromachines* **2021**, *12*, 221. [CrossRef] [PubMed]
12. Zhang, J.; Li, J.; Che, X.; Zhang, X.; Hu, C.; Feng, K.; Xu, T. The Optimal Design of Modulation Angular Rate for MEMS-Based Rotary Semi-SINS. *Micromachines* **2019**, *10*, 111. [CrossRef] [PubMed]
13. EN 50131-2-6. *Alarm Systems. Intrusion and Hol-Up Systems. Part 2-6: Opening Contacts (Magnetic)*; Czech equivalent available on the basis of a license for the Faculty of Security Engineering; CENELEC: Brussels, Belgium, 2009.
14. Zheng, G.; Xue, W.; Chen, H.; Zhang, X.; Hu, C.; Feng, K.; Xu, T. Measurement and Time Response of Electrohydrodynamic Direct-Writing Current. *Micromachines* **2019**, *10*, 90. [CrossRef] [PubMed]
15. Riaz, A.; Noor-ul-Amin, M.; Dogu, E. Effect of measurement error on joint monitoring of process mean and coefficient of variation. *Commun. Stat. Theory Methods* **2021**, 50. [CrossRef]

16. Russenschuck, S.; Caiafa, G.; Fiscarelli, L.; Liebsch, M.; Petrone, C.; Rogacki, P. Challenges in extracting pseudo-multipoles from magnetic measurements. *Int. J. Mod. Phys. A* **2019**, *34*, 1942022. [CrossRef]
17. Kutaj, M.; Boros, M. Development of a new generation of magnetic contact based on hall-effect sensor. In Proceedings of the International Conference of Central-Bohemia-University (CBUIC)—Innovations in Science and Education, Prague, Czech Republic, 22–24 March 2017.
18. Gong, L.L.; Li, X.Y.; Li, Y.H.; Sun, Y.-S.; Lu, H.-H.; Gong, K.-Y.; Guo, Q.; Sun, S.-C.; Li, Z.-Q.; Chen, W. Single-block measurement for the cryogenic permanent magnet undulator sorting. *Radiat. Detect. Technol. Methods* **2020**. [CrossRef]
19. Caruso, M.; Di Tommaso, A.O.; Lisciandrello, G.; Mastromauro, R.A.; Miceli, R.; Nevoloso, C.; Spataro, C.; Trapanese, M. A General and Accurate Measurement Procedure for the Detection of Power Losses Variations in Permanent Magnet Synchronous Motor Drives. *Energies* **2020**, *13*, 5770. [CrossRef]
20. Gavulova, A.; Pirnik, R.; Hudec, R. Technical Support of Traffic Control System of Slovak Agglomerations in NaTIS Project. In Proceedings of the 11th International Conference on Transport Systems Telematics, Katowice Ustron, Poland, 19–22 October 2011; Volume 239.
21. Siser, A.; Lovecek, T.; Maris, L. Simulation of possible assault vectors in an attack using a real-life waterworks object as a use case. In Proceedings of the International Scientific Conference of Young Scientists on Sustainable, Modern and Safe Transport, High Tatras, Slovakia, 31 May–2 June 2017. [CrossRef]
22. Holla, K.; Moricova, V. Specifics of Monitoring and Analysing Emergencies in Information Systems. In Proceedings of the International Scientific Conference of Young Scientists on Sustainable, Modern and Safe Transport, Novy Smokovec, Slovakia, 29–31 May 2019. [CrossRef]
23. Kutaj, M. Probability of Intruder Detection by Selected Components of Intruder alarm System. Ph.D. Thesis, University of Zilina, Zilina, Slovakia, 2017.
24. Kucera, M.; Jarina, R.; Brncal, P.; Brnčal, P.; Gutten, M. Visualisation and Measurement of Acoustic Emission from Power Transformers. In Proceedings of the 12th International Conference on Measurement, Smolenice, Slovakia, 27–29 May 2019.
25. Sebok, M.; Kucera, M.; Gutten, M.; Korenciak, D.; Kubis, M. Non-Destructive Measurement for High-Voltage Transformer of Ignition System. In Proceedings of the 12th International Conference on Measurement, Smolenice, Slovakia, 27–29 May 2019.
26. Boros, M.; Velas, A.; Lenko, F. Utilization of programmable microcontrollers to assess the reliability of alarm systems. In Proceedings of the International Conference on Diagnostics in Electrical Engineering, Pilsen, Czech Republic, 1–4 September 2020.
27. Primin, M.A.; Nedayvoda, I.V. Non-Contact Analysis of Magnetic Fields of Biological Objects: Algorithms for Data Recording and Processing. *Cybern. Syst. Anal.* **2020**, *56*, 848–862. [CrossRef]
28. Mishra, B.; Smirnova, I. Optimal configuration of intrusion detection systems. *Inf. Technol. Manag.* **2021**. [CrossRef]
29. Zhang, R.; Lu, J.Y.; Liu, M.J.; Wang, C.T.; Wang, X.; Gao, E.; Lü, J.; Wang, C.; Kan, F.; Wang, B. Research on Electrical Technology Course Reform based on Professional Certification of Engineering Education. In Proceedings of the International Conference on Education, Management, Computer and Society, Shenyang, China, 1–3 January 2016.
30. Soltes, V.; Stofkova, K.R.; Lenko, F. Socio-economic consequences of globalization on the economic development of regions in the context of security. In Proceedings of the International Scientific Conference Globalization and Its Socio- Economic Consequences—Sustainability in the Global-Knowledge Economy, Rajecke Teplice, Slovakia, 9–10 October 2020.
31. Rehak, D.; Hromada, M.; Lovecek, T. Personnel threats in the electric power critical infrastructure sector and their effect on dependent sectors: Overview in the Czech Republic. *Saf. Sci.* **2020**, *127*, 104698. [CrossRef]
32. Kampova, K.; Makka, K.; Zvarikova, K. Cost benefit analysis within organization security management. In Proceedings of the International Scientific Conference Globalization and Its Socio- Economic Consequences—Sustainability in the Global-Knowledge Economy, Rajecke Teplice, Slovakia, 9–10 October 2020.
33. Rehak, D.; Senovsky, P.; Hromada, M.; Lovecek, T.; Novotny, P. Cascading Impact Assessment in a Critical Infrastructure System. *Int. J. Crit. Infrastruct. Prot.* **2018**, *22*, 125–138. [CrossRef]

 *micromachines*

Article

Free Layer Thickness Dependence of the Stability in $Co_2(Mn_{0.6}Fe_{0.4})Ge$ Heusler Based CPP-GMR Read Sensor for Areal Density of 1 Tb/in^2

Pirat Khunkitti [1,*], Apirat Siritaratiwat [1] and Kotchakorn Pituso [2]

[1] KKU-Seagate Cooperation Research Laboratory, Department of Electrical Engineering, Faculty of Engineering, Khon Kaen University, Khon Kaen 40002, Thailand; apirat@kku.ac.th
[2] Seagate Technology (Thailand) Ltd., Muang 10270, Thailand; k.pituso@gmail.com
* Correspondence: piratkh@kku.ac.th; Tel.: +66-86-636-5678

Abstract: Current-perpendicular-to-the-plane giant magnetoresistance (CPP-GMR) read sensors based on Heusler alloys are promising candidates for ultrahigh areal densities of magnetic data storage technology. In particular, the thickness of reader structures is one of the key factors for the development of practical CPP-GMR sensors. In this research, we studied the dependence of the free layer thickness on the stability of the $Co_2(Mn_{0.6}Fe_{0.4})Ge$ Heusler-based CPP-GMR read head for an areal density of 1 Tb/in^2, aiming to determine the appropriate layer thickness. The evaluations were done through simulations based on micromagnetic modelling. The reader stability indicators, including the magnetoresistance (MR) ratio, readback signal, dibit response asymmetry parameter, and power spectral density profile, were characterized and discussed. Our analysis demonstrates that the reader with a free layer thickness of 3 nm indicates the best stability performance for this particular head. A reasonably large MR ratio of 26% was obtained by the reader having this suitable layer thickness. The findings can be utilized to improve the design of the CPP-GMR reader for use in ultrahigh magnetic recording densities.

Keywords: magnetic recording; magnetic read heads; current perpendicular-to-the-plane giant magnetoresistance; Heusler alloys

Citation: Khunkitti, P.; Siritaratiwat, A.; Pituso, K. Free Layer Thickness Dependence of the Stability in $Co_2(Mn_{0.6}Fe_{0.4})Ge$ Heusler Based CPP-GMR Read Sensor for Areal Density of 1 Tb/in^2. *Micromachines* **2021**, *12*, 1010. https://doi.org/10.3390/mi12091010

Academic Editors: Viktor Sverdlov and Nuttachai Jutong

Received: 9 August 2021
Accepted: 24 August 2021
Published: 25 August 2021

1. Introduction

As is widely claimed by several studies, current-perpendicular-to-the-plane giant magnetoresistance (CPP-GMR) devices have been promising candidates as the magnetic read sensors for ultrahigh areal densities (ADs) of magnetic data storage technology [1–6]. In the past few decades, the outstanding features of the CPP-GMR reader, i.e., large magnetoresistive (MR) outputs, extremely low resistance area (RA) product, capability of transferring large amounts of data at high speeds, and low thermal fluctuation, have been extensively proved [2,7–9]. The very low RA product of the CPP-GMR devices is a key factor in achieving significantly higher ADs than the tunnel magnetoresistance (TMR) junctions used in the current situation [10–13]. The CPP-GMR sensors based on ferromagnetic Heusler alloys are the most capable integrations that can provide very high performance of CPP-GMR sensors nowadays [14–19]. Therefore, several studies have attempted to improve the MR output of the CPP-GMR sensors using various Heusler alloy compositions; however, recent studies indicate that using the $Co_2(Mn_{0.6}Fe_{0.4})Ge$ (CMFG) Heusler alloy as the sensing layer electrodes could provide the highest MR output [20–23].

At ultrahigh ADs in which the media bits must be rapidly downsized, the physical size of the reader needs to be reduced to prevent intertrack interference while maintaining adequate resolution [1,24–26]. The reader shield-to-shield spacing (SSS) is one of the structural parameters directly related to the physical dimension of the head. It also has a major impact on the down-track resolution. Therefore, reducing the SSS is a crucial point for increasing the AD. At an AD of 1 Tb/in^2, the SSS was expected to be less than 25 nm [26]. In particular, the thickness of reader layers is a relative sizing parameter of the SSS. A few nanometers of layers embedded in reader structures are typically desired for the development of practical CPP-GMR read sensors, especially at higher ADs. It is well-known that the thickness of the reader layers also has an influential impact on head performance, particularly the stability of head response [14,27,28]. Thus, the suitable thickness of the head layers should be precisely designed for each one.

In this work, we studied the dependence of the free layer thickness of the CPP-GMR reader on the head's stability performance. The CPP-GMR head based on the CMFG Heusler alloy was focused, assuming that the head was targeted for AD of 1 Tb/in^2. Micromagnetic simulations were based on the finite element method using the M3 code [29]. The rest of this paper is arranged as follows: CPP-GMR modelling is shown in Section 2. Section 3 describes the analysis of read head response. The simulation results, including the related discussions, are given in Section 4. Finally, the results are concluded in Section 5.

2. CPP-GMR Modelling

As shown in Figure 1a, the sensing layers of the CPP-GMR read head targeted for AD of 1 Tb/in^2 are modelled, assuming that the head is sensing the magnetic stray field, $\boldsymbol{H}_{\text{stray}}$, of the medium. The reader width and stripe height of the head were set at 60 and 48 nm, respectively, since this dimension was claimed as the appropriate value for the CPP-GMR reader at an AD of 1 Tb/in^2 [30]. The combination of the CMFG electrodes and AgSn/InZnO spacer was performed due to their suitability for practical CPP-GMR devices [15]. The thickness of the bottom reference layer was 5 nm, while that of the spacer was 2.1 nm. The free layer thickness, t_{FL}, was the main variable in this study, it was varied from 1 to 10 nm. The magnetization of the free layer, $\boldsymbol{M}_{\text{free}}$, was along its easy axis (+y-axis), while the magnetization of the reference layer, $\boldsymbol{M}_{\text{ref}}$, was fixed along the +$x$-axis, assuming that it is due to the exchange bias effect of the anti-ferromagnetic layer. The $\boldsymbol{H}_{\text{stray}}$ produced by the medium was applied to the head on the air bearing surface (x-axis) to mimic the reading situation. The hard bias field, $\boldsymbol{H}_{\text{B}}$, was uniformly supplied to the reader to provide a $\pm 30°$ tilted angle of the free layer magnetization while receiving the $\boldsymbol{H}_{\text{stray}}$. The magnetic media was assumed to be a perpendicular medium having a 10 nm hard layer, while a bit aspect ratio of 4 was set. The medium was based on FePt since it has been widely claimed as a promising material for overcoming the thermal stability limitation at high recording capacities [31,32]. The cross-track magnetic bits are shown in Figure 1b. Their sequence was generated by the 63 pseudorandom bit sequence (PRBS) using the $x^6 + x^5 + 1$ generator polynomial [33]. The gray- and white-filled bits indicate the direction of $\boldsymbol{H}_{\text{stray}}$ along the +x and $-x$ axis, respectively. It is noted that there are no writing errors or intertrack interference included in the simulations, therefore there is no transition noise. The head was assumed to be operated at 1 GHz for practical reasons.

The magnetic properties of CMFG Heusler alloy are adopted from reference [15], as follows: saturation magnetization of 10×10^5 A/m, anisotropy constant of 8×10^3 J/m^3, spin polarization factor of 0.76, Gilbert damping parameter of 0.01, and exchange stiffness constant of 2.25×10^{-11} J/m. The RA product of the sensing layers was 0.11 Ωμm^2. The head was biased with the bias current density of 1.96×10^6 A/cm^2, where its magnitude was purposely limited in order to minimize the influence of spin torque induced instabilities from this current. A positive sign of bias current is when it flows from the reference to the free layers. The device was assumed to be operated at room temperature. The time-varying magnetization was described using the Landau–Lifshitz–Gilbert–Slonczewski (LLGS) formula, as expressed in references [34,35]. A computational cell size of $2.5 \times 2.5 \times 2.5$ nm^3 and a time step of 0.1 ps were set in the simulations.

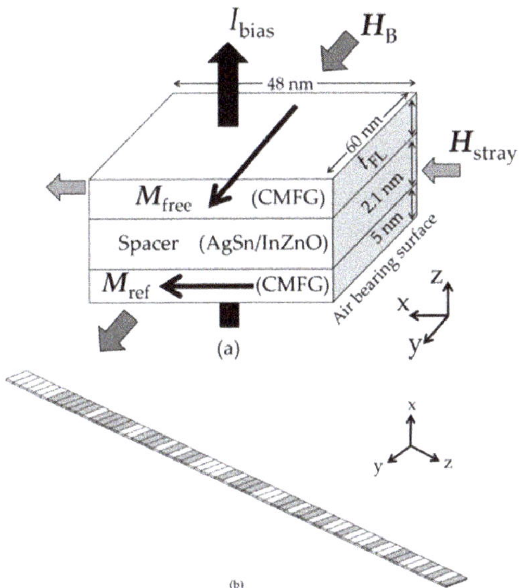

Figure 1. (**a**) CPP-GMR model and (**b**) 63 randomly generated media bits.

3. Analysis of Read Head Response

It is well known that a few nanometers of thickness layers typically have an influential impact on the read head performance, especially at higher ADs. Therefore, the layer thickness of Heusler alloy films embedded in the CPP-GMR structure needs to be optimally designed to achieve the desired physical dimension of practical CPP-GMR sensors. To investigate the dependence of free layer thickness on the CPP-GMR reader's stability performance, the output characteristics of the head, including the MR ratio, readback signal, dibit response, asymmetry parameter, and power spectral density (PSD) profile, were analyzed and discussed. The MR ratio basically represents the amplitude of the sensor's output. The readback signal typically indicates the head response. It is obtained from the magnetization dynamic of the free layer passing through the Butterworth low-pass filter [36]. Based on the readback signal pattern, the dibit response is another important parameter indicating the nonlinear behavior and distortion occurring in the readback waveform. In this work, we performed the domain dibit extraction technique to obtain the linear dibit response as well as the nonlinearities via echoes around the main pulses [37]. In addition, an asymmetry parameter can be calculated from the difference between the positive and negative readback amplitudes, as written in Equation (1) [38].

$$\% Asymmetry = \frac{(Vp - Vn)}{(Vp + Vn)} \times 100 \tag{1}$$

The PSD profile demonstrates the fluctuation of the time-varying magnetization, as well as indicates the frequency spectrum of the readback signal. The local PSD is firstly calculated through the time-varying magnetization, $\boldsymbol{M}_{x,y,z}(\boldsymbol{r}_i,t_j)$, where $\boldsymbol{r}_i$ is the magnetization position at each varying time, t_j, given in Equation (2) [39].

$$S_{x,y,z}(r_i, f) = \left| \sum_j \boldsymbol{M}_{x,y,z}(\boldsymbol{r}_i, t_j) e^{i2\pi f t_j} \right|^2 \tag{2}$$

Then, the total PSD was computed by a summation of the local PSD at each particular frequency, $S_{x,y,z}(\mathbf{r}_i, f)$, given as Equation (3). An integrated PSD can be further obtained by an integral of the overall PSD.

$$\overline{S}_{x,y,z}(f) = \sum_i S_{x,y,z}(\mathbf{r}_i, f) \tag{3}$$

4. Results and Discussion

In this section, the output characteristics of the CPP-GMR read head were characterized at different t_{FL} from 1 to 10 nm. The variation range of t_{FL} was based on the possible scale for practical devices while taking the covering trend of results into account. The focused parameters used for indicating the head stability performance, including the MR ratio, readback signal, dibit response, asymmetry parameter, and PSD profile, were analyzed and discussed.

The MR ratio of the CPP-GMR reader versus t_{FL} is presented in Figure 2. It shows that the MR ratio increases at thicker t_{FL}. Above a t_{FL} of 6 nm, a change in layer thickness has less impact on the MR ratio increment than below. An enhancement of bulk spin-dependent scattering contributed to an increase in MR ratio.

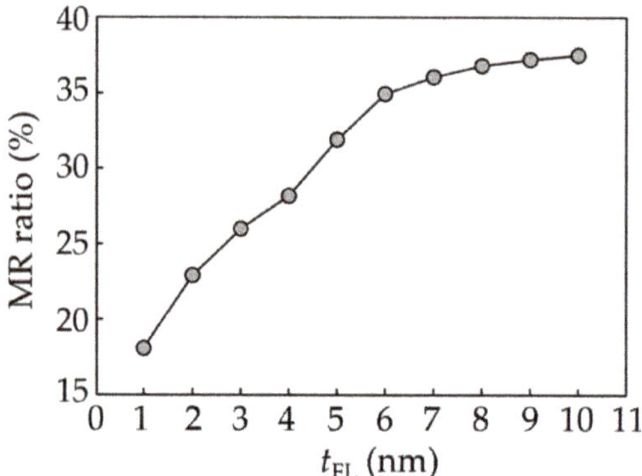

Figure 2. MR ratio of the CPP-GMR reader at various free layer thicknesses.

To characterize the readback response of the CPP-GMR reader, we investigated the readback signal of the reader at t_{FL} of 1 to 10 nm. Examples of readback signal waveforms for t_{FL} of 1, 3, 5, and 8 nm are illustrated in Figure 3. The reader with a t_{FL} of 8 nm appears to have the highest distortion in the readback signal waveform. Meanwhile, the readback signals of the readers with t_{FL} of 3 and 5 nm are well patterned and symmetric. However, it is generally insufficient to analyze the readback response through only an investigation of the readback waveform. We therefore characterized more insights related to the readback signal behavior, which are the dibit response, asymmetry parameter, and the PSD profile. These parameters usually correspond to the stability performance of the read sensors.

The dibit response of the readback signal was obtained though the 63-bit PRBS with the polynomial $x^6 + x^5 + 1$. The main echoes related to this response are $C^{(2)}{}_1$ and $C^{(2)}{}_2$, which are located at bits 27 and 22, respectively. These echoes typically indicate the non-linear distortion of the readback waveform due to reader asymmetry [37]. It is noted that the impacts of higher orders of echoes were dominated by these main echoes and can be neglected in the evaluations. Figure 4a demonstrates the examples of dibit extraction of the reader with t_{FL} of 1, 3, 5, and 8 nm. Each echo is magnified in the insets. The readers having a t_{FL} of 5 and 8 nm seem to have higher echo amplitudes than others. Further details of echoes' amplitudes were, in addition, analyzed at all possible t_{FL}, as shown in Figure 4b. When the t_{FL} was increased starting from 1 nm, the amplitude of the echoes decreased until it reached its lowest scale. Then, the amplitude of echoes increased continuously when the

t_{FL} increased beyond the point providing the lowest echo amplitude. A variation of t_{FL} appears to have a minor impact on $C^{(2)}{}_1$ and $C^{(2)}{}_2$ at a t_{FL} above 6 nm, indicating a lesser affectation on the readback signal distortion for this specific range.

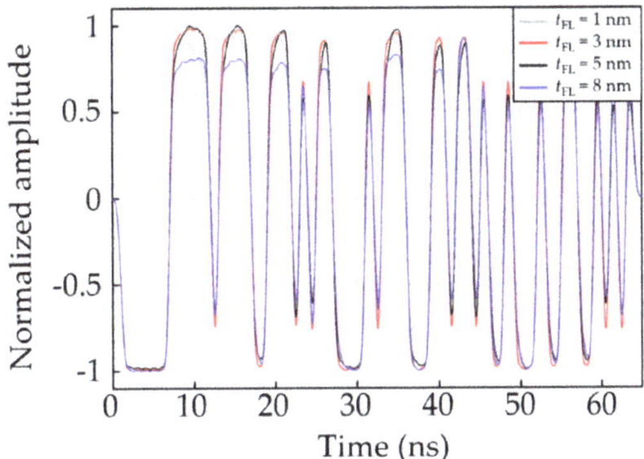

Figure 3. Examples of readback signal of the CPP-GMR reader.

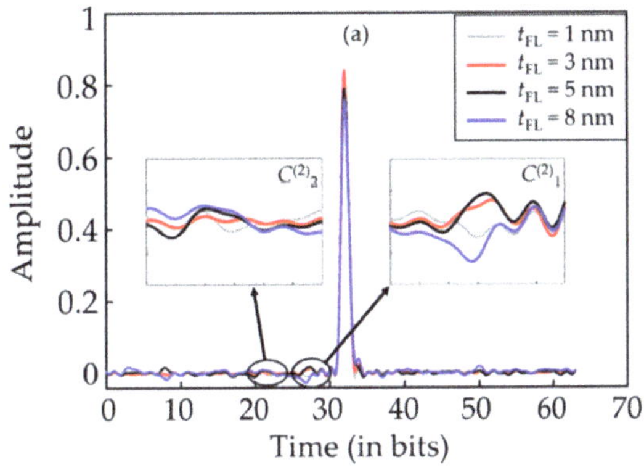

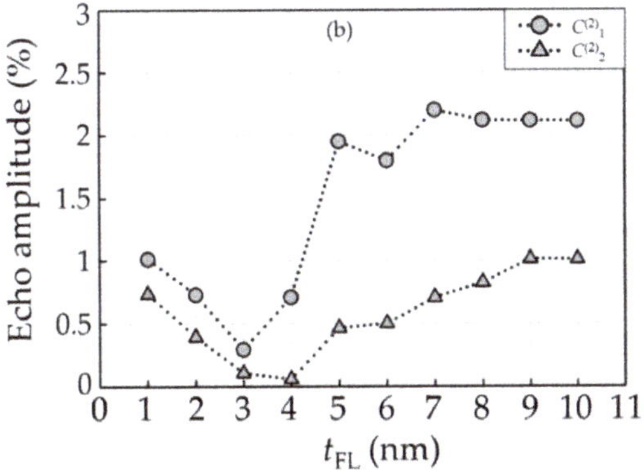

Figure 4. Dibit extraction of the CPP-GMR reader with different free layer thicknesses; (**a**) examples of dibit response; (**b**) amplitude of echoes.

The asymmetry parameter of the readback signal obtained from the CPP-GMR reader was examined at different t_{FL}, as shown in Figure 5. Most readers, except for those with a t_{FL} of 6 and 7 nm, were found to contain under 10% readback signal asymmetry. The readback waveforms produced by the readers with a t_{FL} of 4 and 9 nm are the most symmetric. From analysis of the readback signal, its dibit extraction, and its asymmetry parameter, it is obviously seen that the most suitable thickness of the free layer is 3 nm. The reader with this thickness value could provide the greatest pattern of readback signal.

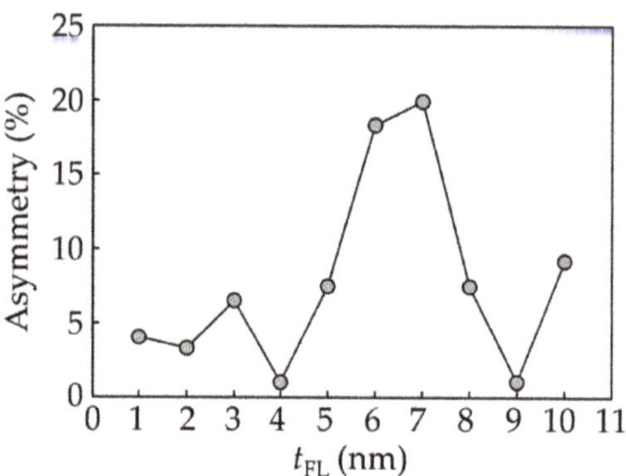

Figure 5. Asymmetry parameter obtained from the CPP-GMR reader's response.

In addition, the PSD profile of the head response was characterized as another stability indicator for the reader. Figure 6a illustrates the integrated PSD at various t_{FL}, while its frequency spectrum is indicated in Figure 6b. The PSD scale was found to be continuously lowered as the t_{FL} was reduced from 10 to 3 nm. Below t_{FL} = 3 nm, an adjustment of the t_{FL} causes a slight change in PSD amplitude. Then, it is worth reducing the thickness of the free layer to 3 nm. The frequency spectrum of PSD of the readers with t_{FL} of 1, 3, 5, and 8 nm is demonstrated in Figure 6b. Corresponding to Figure 6a, the amplitude of PSD becomes smaller at greater t_{FL}. In particular, we found that the spectral peak is shifted to higher frequencies by decreasing the t_{FL}. This frequency shifting behavior can be described by the magnetization precession which is computed by the LLGS equation [33,34]. The LLGS formula generally consists of the precession, damping, and spin torque terms. The spectral peak theoretically occurs depending mainly on the precession and damping terms of the time-varying magnetization. As the t_{FL} is reduced, the spin torque term in which its direction is opposing the magnetization precession becomes higher. This accordingly causes an enhancement of the force pulling the magnetization towards the opposite direction to its initial state. The resulting force therefore yields the higher oscillation of the magnetization precession. Rather than the frequency shifting, the higher PSD intensity at greater t_{FL} indicates a stronger impact on the reader stability, as this typically implies less stable magnetization precession and may further reduce the signal-to-noise ratio of the read sensors.

In summary, as the narrower physical reader gap of the read sensors is required to reach higher ADs, reducing the thickness of the free layer is therefore another approach to achieve this requirement. A very thin free layer, on the other hand, may result in an insufficient MR ratio. Then, in order to provide effective signal processing, an adequate MR ratio must be maintained. Our analysis shows that although a thinner free layer could provide a better readback response, the MR ratio is also reduced. Based on the trade-off between all characterized parameters, we believe that the appropriate t_{FL} of this particular CPP-GMR reader is 3 nm. At this point, a reasonable MR ratio of 26% is sufficient for practical devices. The highest stability performance of the reader with a t_{FL} of 3 nm was also confirmed through the analysis of the readback response that it is worthwhile to reduce the t_{FL} to 3 nm.

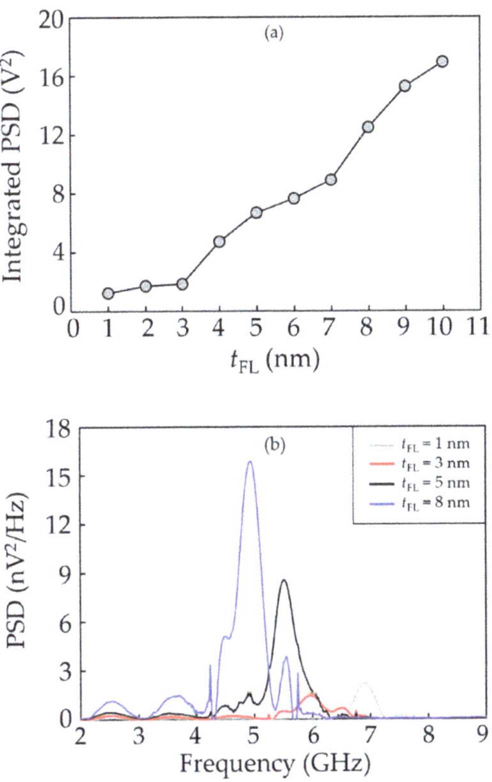

Figure 6. The PSD profile of the CPP-GMR reader at various free layer thicknesses; (**a**) integrated PSD; (**b**) frequency spectrum.

5. Conclusions

In this work, we investigated the dependence of the t_{FL} on the stability performance of the CMFG Heusler-based CPP-GMR sensor targeted for an areal density of 1 Tb/in^2. Simulations were done based on micromagnetic modelling. It was found that the t_{FL} has a highly influential impact on the MR ratio at t_{FL} below 6 nm. A consideration of the readback signal of the head, including its dibit extraction and asymmetry parameter, indicates that the reader having a t_{FL} of 3 nm could produce a greatly patterned readback waveform. The PSD profile and its frequency spectrum are, in addition, analyzed and discussed to confirm the worthiness of setting the t_{FL} to 3 nm. Results also showed that a reasonably large MR ratio of 26% was greatly maintained at a t_{FL} of 3 nm. Therefore, the trade-off between all evaluated parameters suggests that this particular CPP-GMR reader with a t_{FL} of 3 nm indicates the best stability performance. Findings can be utilized to design the CPP-GMR reader for use in ultrahigh areal densities of magnetic data storage.

Author Contributions: Conceptualization, P.K.; data curation, P.K.; methodology, P.K. and A.S.; software, P.K.; formal analysis, P.K., A.S. and K.P.; investigation, P.K.; validation, P.K.; writing—original draft, P.K.; writing—Review and editing, P.K.; visualization, A.S. and K.P.; funding acquisition, P.K. All authors have read and agreed to the published version of the manuscript.

Funding: This work was financially supported by Office of the Permanent Secretary, Ministry of Higher Education, Science, Research and Innovation, Thailand (Grant No. RGNS 63-058).

Institutional Review Board Statement: Not applicable.

Informed Consent Statement: Not applicable.

Conflicts of Interest: The authors declare no conflict of interest.

References

1. Takagishi, M.; Yamada, K.; Iwasaki, H.; Fuke, H.N.; Hashimoto, S. Magnetoresistance Ratio and Resistance Area Design of CPP-MR Film for 2–5 Tb/in^2 Read Sensors. *IEEE Trans. Magn.* **2010**, *46*, 2086–2089. [CrossRef]
2. Nagasaka, K. CPP-GMR technology for magnetic read heads of future high-density recording systems. *J. Magn. Magn. Mater.* **2009**, *321*, 508–511. [CrossRef]
3. Childress, J.R.; Fontana, R.E. Magnetic recording read head sensor technology. *Comptes Rendus Phys.* **2005**, *6*, 997–1012. [CrossRef]
4. Nakatani, T.; Gao, Z.; Hono, K. Read sensor technology for ultrahigh density magnetic recording. *MRS Bull.* **2018**, *43*, 106–111. [CrossRef]
5. Khunkitti, P.; Siritaratiwat, A.; Kaewrawang, A.; Mewes, T.; Mewes, C.; Kruesubthaworn, A. Electromagnetic interference-induced instability in CPP-GMR read heads. *J. Magn. Magn. Mater.* **2016**, *412*, 42–48. [CrossRef]
6. Pipathanapoompron, T.; Stankiewicz, A.; Wang, J.; Subramanian, K.; Kaewrawang, A. Magnetic reader testing for asymmetric oscillation noise. *J. Magn. Magn. Mater.* **2020**, *514*, 167064. [CrossRef]
7. Takagishi, M.; Koi, K.; Yoshikawa, M.; Funayama, T.; Iwasaki, H.; Sahashi, M. The applicability of CPP-GMR heads for magnetic recording. *IEEE Trans. Magn.* **2002**, *38*, 2277–2282. [CrossRef]
8. Diao, Z.; Chapline, M.; Zheng, Y.; Kaiser, C.; Roy, A.G.; Chien, C.; Shang, C.; Ding, Y.; Yang, C.; Mauri, D.; et al. Half-metal CPP GMR sensor for magnetic recording. *J. Magn. Magn. Mater.* **2013**, *356*, 73–81. [CrossRef]
9. Childress, J.R.; Carey, M.J.; Maat, S.; Smith, N.; Fontana, R.E.; Druist, D.; Carey, K.; Katine, J.A.; Robertson, N.; Boone, T.D.; et al. All-Metal Current-Perpendicular-to-Plane Giant Magnetoresistance Sensors for Narrow-Track Magnetic Recording. *IEEE Trans. Magn.* **2007**, *44*, 90–94. [CrossRef]
10. Surawanitkun, C.; Kaewrawang, A.; Siritaratiwat, A.; Kruesubthaworn, A.; Sivaratana, R.; Jutong, N.; Mewes, C.K.A.; Mewes, T. Magnetic Instability in Tunneling Magnetoresistive Heads Due to Temperature Increase During Electrostatic Discharge. *IEEE Trans. Device Mater. Reliab.* **2012**, *12*, 570–575. [CrossRef]
11. Teso, B.; Kravenkit, S.; Sorn-In, K.; Kaewrawang, A.; Kruesubthaworn, A.; Siritaratiwat, A.; Mewes, T.; Mewes, C.; Surawanitkun, C. Temperature dependence of magnetic properties on switching energy in magnetic tunnel junction devices with tilted magnetization. *Appl. Surf. Sci.* **2019**, *472*, 36–39. [CrossRef]
12. Surawanitkun, C.; Kaewrawang, A.; Siritaratiwat, A.; Kruesubthaworn, A.; Sivaratana, R.; Jutong, N.; Mewes, C.; Mewes, T. Modeling of switching energy of magnetic tunnel junction devices with tilted magnetization. *J. Magn. Magn. Mater.* **2015**, *381*, 220–225. [CrossRef]
13. Teso, B.; Siritaratiwat, A.; Kaewrawang, A.; Kruesubthaworn, A.; Namvong, A.; Sainon, S.; Surawanitkun, C. Switching Performance Comparison with Low Switching Energy Due to Initial Temperature Increment in CoFeB/MgO-Based Single and Double Barriers. *IEEE Trans. Electron Devices* **2019**, *66*, 4062–4067. [CrossRef]
14. Nakatani, T.M.; Hase, N.; Goripati, H.S.; Takahashi, Y.K.; Furubayashi, T.; Hono, K. Co-Based Heusler Alloys for CPP-GMR Spin-Valves with Large Magnetoresistive Outputs. *IEEE Trans. Magn.* **2012**, *48*, 1751–1757. [CrossRef]
15. Nakatani, T.; Li, S.; Sakuraba, Y.; Furubayashi, T.; Hono, K. Advanced CPP-GMR Spin-Valve Sensors for Narrow Reader Applications. *IEEE Trans. Magn.* **2017**, *54*, 1–11. [CrossRef]
16. Wen, Z.; Kubota, T.; Ina, Y.; Takanashi, K. Dual-spacer nanojunctions exhibiting large current-perpendicular-to-plane giant magnetoresistance for ultrahigh density magnetic recording. *Appl. Phys. Lett.* **2017**, *110*, 102401. [CrossRef]
17. Du, Y.; Furubayashi, T.; Sasaki, T.T.; Sakuraba, Y.; Takahashi, Y.K.; Hono, K. Large magnetoresistance in current-perpendicular-to-plane pseudo spin-valves using $Co_2Fe(Ga_{0.5}Ge_{0.5})$ Heusler alloy and AgZn spacer. *Appl. Phys. Lett.* **2015**, *107*, 112405. [CrossRef]
18. Pradines, B.; Calmels, L.; Arras, R. Robustness of the Half-Metallicity at the Interfaces in Co2MnSi -Based All-Full-Heusler-Alloy Spintronic Devices. *Phys. Rev. Appl.* **2021**, *15*, 034009. [CrossRef]
19. Kubota, T.; Wen, Z.; Takanashi, K. Current-perpendicular-to-plane giant magnetoresistance effects using Heusler alloys. *J. Magn. Magn. Mater.* **2019**, *492*, 165667. [CrossRef]
20. Nakatani, T.; Sasaki, T.T.; Sakuraba, Y.; Hono, K. Improved current-perpendicular-to-plane giant magnetoresistance outputs by heterogeneous Ag-In:Mn-Zn-O nanocomposite spacer layer prepared from Ag-In-Zn-O precursor. *J. Appl. Phys.* **2019**, *126*, 173904. [CrossRef]
21. Page, M.R.; Nakatani, T.M.; Stewart, D.A.; York, B.R.; Read, J.C.; Choi, Y.-S.; Childress, J.R. Temperature-dependence of current-perpendicular-to-the-plane giant magnetoresistance spin-valves using $Co_2(Mn_{1-x}Fe_x)Ge$ Heusler alloys. *J. Appl. Phys.* **2016**, *119*, 153903. [CrossRef]
22. Li, S.; Nakatani, T.; Masuda, K.; Sakuraba, Y.; Xu, X.; Sasaki, T.; Tajiri, H.; Miura, Y.; Furubayashi, T.; Hono, K. Enhancement of current-perpendicular-to-plane giant magnetoresistive outputs by improving B2-order in polycrystalline Co2(Mn0.6Fe0.4)Ge Heusler alloy films with the insertion of amorphous CoFeBTa underlayer. *Acta Mater.* **2018**, *142*, 49–57. [CrossRef]
23. Nakatani, T.; Sasaki, T.T.; Li, S.; Sakuraba, Y.; Furubayashi, T.; Hono, K. The microstructural origin of the enhanced current-perpendicular-to-the-plane giant magnetoresistance by Ag/In-Zn-O/Zn spacer layer. *J. Appl. Phys.* **2018**, *124*, 223904. [CrossRef]

24. Han, G.C.; Qiu, J.J.; Wang, L.; Yeo, W.K.; Wang, C.C. Perspectives of Read Head Technology for 10 Tb/in^2 Recording. *IEEE Trans. Magn.* **2010**, *46*, 709–714. [CrossRef]
25. Stankiewicz, A.; Pipathanapoompron, T.; Subramanian, K.; Kaewrawang, A. Reader Noise Due to Thermally Driven Asymmetric Oscillations. *IEEE Trans. Magn.* **2018**, *54*, 1–5. [CrossRef]
26. Khunkitti, P.; Kaewrawang, A.; Siritaratiwat, A.; Mewes, T.; Mewes, C.K.A.; Kruesubthaworn, A. A novel technique to detect effects of electromagnetic interference by electrostatic discharge simulator to test parameters of tunneling magnetoresistive read heads. *J. Appl. Phys.* **2015**, *117*, 17A908. [CrossRef]
27. Kubota, T.; Ina, Y.; Wen, Z.; Narisawa, H.; Takanashi, K. Current perpendicular-to-plane giant magnetoresistance using an L12 Ag3 Mg spacer and Co2 Fe0.4 Mn0.6 Si Heusler alloy electrodes: Spacer thickness and annealing temperature dependence. *Phys. Rev. Mater.* **2017**, *1*, 4. [CrossRef]
28. Nakatani, T.; Narayananellore, S.K.; Kumara, L.S.R.; Tajiri, H.; Sakuraba, Y.; Hono, K. Thickness dependence of degree of B2 order of polycrystalline Co2(Mn0.6Fe0.4)Ge Heusler alloy films measured by anomalous X-ray diffraction and its impacts on current-perpendicular-to-plane giant magnetoresistance properties. *Scr. Mater.* **2020**, *189*, 63–66. [CrossRef]
29. Mewes, T.; Mewes, C.K.A. Matlab Based Micromagnetics Code M3. 2012. Available online: http://magneticslab.ua.edu/micromagnetics-code.html/ (accessed on 11 April 2021).
30. Khunkitti, P.; Kruesubthaworn, A.; Kaewrawang, A.; Siritaratiwat, A. Unstable Playback Response of CPP-GMR Read Head Induced by Electromagnetic Interference: Structural Dependence. *IEEE Trans. Magn.* **2019**, *55*, 1–6. [CrossRef]
31. Tipcharoen, W.; Kaewrawang, A.; Siritaratiwat, A. Design and Micromagnetic Simulation of Fe/L10-FePt/Fe Trilayer for Exchange Coupled Composite Bit Patterned Media at Ultrahigh Areal Density. *Adv. Mater. Sci. Eng.* **2015**, *2015*, 1–5. [CrossRef]
32. Pituso, K.; Kaewrawang, A.; Buatong, P.; Siritaratiwat, A.; Kruesubthaworn, A. The temperature and electromagnetic field distributions of heat-assisted magnetic recording for bit-patterned media at ultrahigh areal density. *J. Appl. Phys.* **2015**, *117*, 17C501. [CrossRef]
33. Macwilliams, F.J.; Sloane, N.J.A. Pseudo Random Sequences and Arrays. *Proc. IEEE* **1976**, *64*, 1715–1729. [CrossRef]
34. Slonczewski, J. Current-driven excitation of magnetic multilayers. *J. Magn. Magn. Mater.* **1996**, *159*, L1–L7. [CrossRef]
35. Berger, L. Emission of spin waves by a magnetic multilayer traversed by a current. *Phys. Rev. B* **1996**, *54*, 9353–9358. [CrossRef]
36. Kovintavewat, P.; Ozgunes, I.; Kurtas, E.; Barry, J.; McLaughlin, S. Generalized partial-response targets for perpendicular recording with jitter noise. *IEEE Trans. Magn.* **2002**, *38*, 2340–2342. [CrossRef]
37. Eppler, W.; Ozgunes, I. Channel characterization methods using dipulse extraction. *IEEE Trans. Magn.* **2006**, *42*, 176–181. [CrossRef]
38. Van Der Heijden, P.A.A.; Karns, D.W.; Clinton, T.W.; Heinrich, S.J.; Batra, S.; Karns, D.C.; Roscamp, T.A.; Boerner, E.D.; Eppler, W.R. The effect of media background on reading and writing in perpendicular recording. *J. Appl. Phys.* **2002**, *91*, 8372. [CrossRef]
39. McMichael, R.; Stiles, M. Magnetic normal modes of nanoelements. *J. Appl. Phys.* **2005**, *97*, 10J901. [CrossRef]

micromachines

MDPI

Article

Micromagnetic Simulation of $L1_0$-FePt-Based Exchange-Coupled-Composite-Bit-Patterned Media with Microwave-Assisted Magnetic Recording at Ultrahigh Areal Density

Pirat Khunkitti [1], Naruemon Wannawong [2], Chavakon Jongjaihan [1] Apirat Siritaratiwat [1], Anan Kruesubthaworn [1] and Arkom Kaewrawang [1,*]

[1] Department of Electrical Engineering, Faculty of Engineering, Khon Kaen University, Khon Kaen 40002, Thailand; piratkh@kku.ac.th (P.K.); chavakon.j@kkumail.com (C.J.); apirat@kku.ac.th (A.S.); anankr@kku.ac.th (A.K.)

[2] Seagate Technology (Thailand) Co., Ltd., Nakhon Ratchasima 30170, Thailand; yongyangbobby@gmail.com

* Correspondence: arkom@kku.ac.th

Abstract: In this work, we propose exchange-coupled-composite-bit-patterned media (ECC-BPM) with microwave-assisted magnetic recording (MAMR) to improve the writability of the magnetic media at a 4 Tb/in^2 recording density. The suitable values of the applied microwave field's frequency and the exchange coupling between magnetic dots, A_{dot}, of the proposed media were evaluated. It was found that the magnitude of the switching field, H_{sw}, of the bilayer ECC-BPM is significantly lower than that of a conventional BPM. Additionally, using the MAMR enables further reduction of H_{sw} of the ECC-BPM. The suitable frequency of the applied microwave field for the proposed media is 5 GHz. The dependence of A_{dot} on the H_{sw} was additionally examined, showing that the A_{dot} of 0.14 pJ/m is the most suitable value for the proposed bilayer ECC-BPM. The physical explanation of the H_{sw} of the media under a variation of MAMR and A_{dot} was given. Hysteresis loops and the magnetic domain of the media were characterized to provide further details on the results. The lowest H_{sw} found in our proposed media is 12.2 kOe, achieved by the bilayer ECC-BPM with an A_{dot} of 0.14 pJ/m using a 5 GHz MAMR.

Keywords: bit-patterned media; exchange-coupled-composite media; microwave-assisted magnetic recording; hysteresis loop

Citation: Khunkitti, P.; Wannawong, N.; Jongjaihan, C.; Siritaratiwat, A.; Kruesubthaworn, A.; Kaewrawang, A. Micromagnetic Simulation of $L1_0$-FePt-Based Exchange-Coupled-Composite-Bit-Patterned Media with Microwave-Assisted Magnetic Recording at Ultrahigh Areal Density. *Micromachines* **2021**, *12*, 1264. https://doi.org/10.3390/mi12101264

Academic Editors: Viktor Sverdlov and Nuttachai Jutong

Received: 31 August 2021
Accepted: 14 October 2021
Published: 17 October 2021

1. Introduction

Recently, the capacity of hard disk drives has been heading towards a stalemate since the thermal stability of the conventional media has reached its limitation [1–6]. In order to increase recording densities, several techniques have been proposed to overcome the stability limitation, such as exchange-coupled-composite (ECC) media, heat-assisted magnetic recording (HAMR) bit-patterned media (BPM), and microwave-assisted magnetic recording (MAMR) [1,7–14]. The ECC media have been introduced to improve the magnetic properties of the media; the major goal is to reduce the magnitude of the switching field, H_{sw} [7,12,13]. The ECC media consist of magnetically isolated layers, which are the soft and the hard magnetic layers. The interface exchange coupling between two layers can provide a lower H_{sw} than the conventional one. It therefore enables the use of smaller media grain sizes with higher magnetic anisotropy, K_u, at high areal densities [7]. BPM technology has been extensively proposed to solve the magnetic transition noise and the interaction between magnetic bits of the conventional media since these factors could be the crucial issues at ultrahigh areal densities [1]. The principle of BPM is the magnetic separation of each magnetic bit, which eliminates the magnetic transition noise and yields a very low interaction between bits [1]. In previous research, the combination of ECC-BPM was reported, demonstrating that this combined technique can increasingly reduce the H_{sw} of the media beyond that using each technique individually [12]. Microwave-assisted magnetic recording (MAMR) is one promising technology for achieving higher areal densities [14–18]. The goal of this technique is to reduce the H_{sw} of the media by using an AC-magnetic field, H_{ac}, operated at a microwave range. During the writing process of MAMR, the H_{ac} is applied to the media simultaneously with the writing field, $H_{dc.}$ This strategy can increasingly reduce the energy barrier of the media by the ferromagnetic resonance (FMR) phenomenon. As a result, switching the magnetization of the media becomes easier. A significant reduction in the H_{sw} of the media under the use of MAMR can be seen in several recent publications [14–18].

As it is generally known that the maximum write head field is limited, it is challenging to continuously reduce the H_{sw} to enable the use of higher K_u materials as a media at higher recording densities [19]. Therefore, we proposed a novel technique to improve the writability of the media by the combination of three technologies, including the ECC media, BPM, and MAMR. The proposed media was targeted for use at an areal density of 4 Tb/in^2. The hybrid magnetic recording media based on the FePt alloy with the face-centered tetragonal $L1_0$ structure, $L1_0$-FePt, was focused. The object-oriented micro-magnetic framework (OOMMF) software [20] was used in the simulations implementing the Landau–Lifshitz–Gilbert (LLG) equation.

2. Modeling and Analytical Methodology

The model structures of the $L1_0$-FePt-based single-layer BPM and the $L1_0$-FePt/Fe bilayer ECC-BPM for an areal density of 4 Tb/in^2 are shown in Figure 1a,b, respectively. The media configuration has been determined on the basis of BPM and ECC-BPM requirements. To achieve an areal density of 4 Tb/in^2, the cube dot size of single-layer BPM of $10 \times 10 \times 10$ nm^3 and spacing between dots of 2.5 nm were assumed [12]. The bilayer ECC-BPM was introduced by magnetically adding a soft Fe layer with a 10 nm thickness under the FePt BPM. The Fe added layer was assumed to have the same dot pattern as the single-layer BPM. The magnetization of each layer was initially aligned along the +z direction for both media. The magnetic properties of the proposed media are detailed as follows: the $L1_0$-FePt hard layer had a saturation magnetization, M_s, of 1.175 MA/m, and a K_u of 2.8 MJ/m^3. The Fe soft layer had M_s of 1.71 MA/m and K_u of 100 J/m^3. For the bilayer media, the exchange coupling between soft and hard layers, A_{ex}, was assumed to be 25 pJ/m, whereas the exchange coupling between magnetic dots, A_{dot}, varied from 0.1 to 0.3 pJ/m. This variation of A_{dot} was based on the possible values of the filled material between the dot spacing, as can be seen in the literature [21,22]. In calculations, the time-varying magnetization was described by the LLG equation, given as Equation (1) [23]:

$$\frac{\partial \vec{M}}{\partial t} = -|\gamma|\mu_0(\vec{M} \times \vec{H}_{eff} + \frac{\alpha}{M_s}\vec{M} \times (\vec{M} \times \vec{H}_{eff})) \tag{1}$$

where $\vec{M}$ is the magnetization, γ is the gyromagnetic ratio, and $\vec{H}_{eff}$ is the effective magnetic field given by Equation (2):

$$\vec{H}_{eff} = \vec{H}_{dc} + \vec{H}_{ex} + \vec{H}_{de} + \vec{H}_k + h_{ac}\sin(2\pi f_{ac}t)\vec{a}_x \tag{2}$$

where $\vec{H}_{dc}$, $\vec{H}_{ex}$, $\vec{H}_{de}$, $\vec{H}_k$, h_{ac}, f_{ac}, and $\vec{a}_x$ are the external static magnetic field, the exchange field between nearest-neighbor cells, the demagnetizing field, the field due to uniaxial anisotropy, the amplitude of microwave field, the microwave frequency, and the unit vector along the x-axis, respectively.

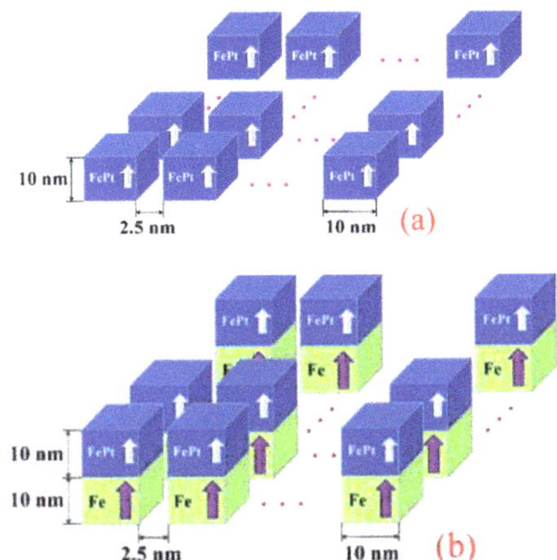

Figure 1. (**a**) Single layer BPM and (**b**) bilayer ECC-BPM.

To investigate the writability of the media, the external DC write field, H_{dc}, was applied to the media in the z-direction. The write head field region was defined as the region where the magnetization of the media can be written by the write head field. The H_{sw} of the media was collected in the condition that $M = -0.8M_s$. For the MAMR included, the H_{ac} was simultaneously applied to H_{dc} with a magnitude of 100 mT in the x-direction. The H_{sw} of the proposed ECC-BPM was examined at various frequencies of H_{ac}. Since A_{dot} typically has an influential impact on the H_{sw}, the dependence of A_{dot} on the H_{sw} was also taken into account. Then, the hysteresis loops and magnetic domain of a bilayer ECC-BPM were determined and compared with those of a single-layer BPM.

3. Results and Discussions

The H_{sw} of the proposed bilayer ECC-BPM at different A_{dot} values was investigated at f_{ac} between 0 and 25 GHz, as shown in Figure 2. It was discovered that changing A_{dot} or f_{ac} could change the H_{sw}. Without MAMR, the media with an A_{dot} of 0.14 pJ/m have the lowest H_{sw}, followed by those with an A_{dot} of 0.12, 0.25, and 0.30 pJ/m, respectively. A variation of A_{dot} under the MAMR indicates a similar trend of H_{sw} for all f_{ac}. When the MAMR was performed, it revealed that increasing the f_{ac} from 0 to 2 GHz slightly changes the H_{sw} of the media. Then, the H_{sw} was dramatically reduced to the range of f_{ac} about 5–7.5 GHz. Above f_{ac} of 10 GHz, the H_{sw} of all media tended to have a higher H_{sw} than that without MAMR and was insignificantly changed with varying f_{ac}. From the results, the media with A_{dot} = 0.25, 0.30 pJ/m indicate their lowest H_{sw} at a f_{ac} of 5 GHz, whereas the lowest H_{sw} of media with A_{dot} = 0.12, 0.14 pJ/m occurs at a f_{ac} of 7.5 GHz. The lowest H_{sw} found in this evaluation is 11.9 kOe, achieved by the media with an A_{dot} of 0.14 pJ/m with MAMR at a f_{ac} of 7.5 GHz. However, since the resonance frequency of our proposed system is 5 GHz, we therefore examined more details of A_{dot}'s influence on the H_{sw} at this frequency.

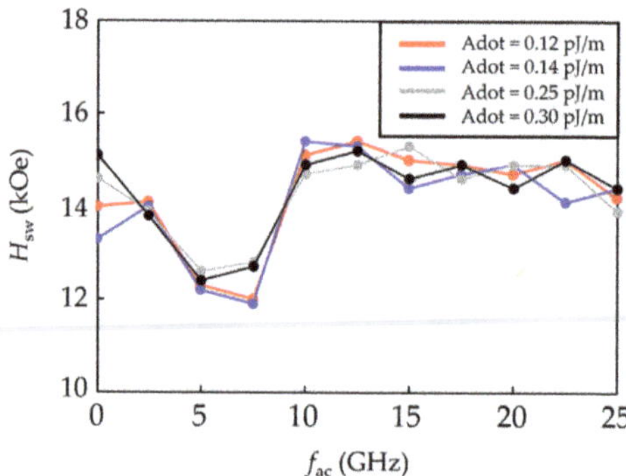

Figure 2. H_{SW} of the bilayer ECC-BPM versus f_{ac} at various A_{dot}.

Figure 3 displays the H_{SW} of the bilayer ECC-BPM as a function of A_{dot}, in the cases without and with the MAMR at f_{ac} = 5 GHz. Overall, it shows that using MAMR can significantly reduce the H_{SW} of the media for all A_{dot} values. Additionally, the media with MAMR indicate less variation of H_{SW} with a varying A_{dot} than that without MAMR. The lowest H_{SW} of ECC-BPM without and with MAMR is 13.1 kOe at A_{dot} = 0.16 pJ/m and 12.2 kOe at A_{dot} = 0.14 pJ/m, respectively. The H_{SW} alternation with varying A_{dot} at other f_{ac} is supposed to provide a higher H_{SW} than that at f_{ac} = 5 GHz due to its larger conventional value and therefore is not considered in this work.

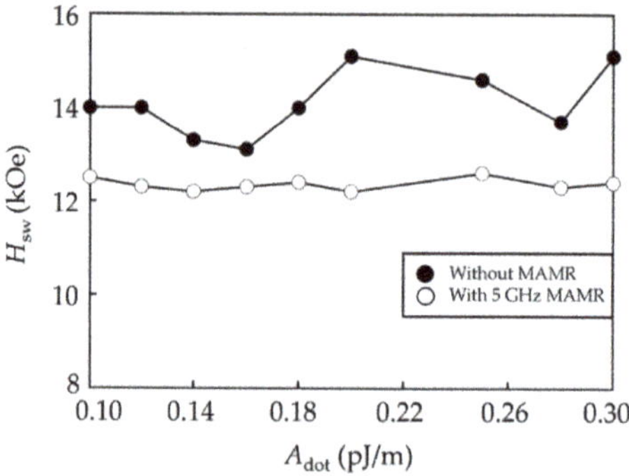

Figure 3. H_{SW} of the bilayer ECC-BPM with and without HAMR as a function of A_{dot}.

In a case without MAMR, the media with higher A_{dot} tend to have higher H_{SW}. A magnitude of A_{dot} normally represents an exchange interaction between magnetic dots of the media. This interaction indicates an interacting force between adjacent magnetizations. At higher A_{dot}, each magnetic dot is strongly related to each other due to a massive exchange interaction between them. When the write head field is applied to the media, the magnetization of the media with higher A_{dot} is more homogeneously processed, then it is more difficult to be switched.

The physical reason why a variation of f_{ac} can alter the H_{SW} of media with MAMR can be explained by the FMR phenomenon, as follows: when the magnetization precession of a ferromagnetic material is under the effect of an external magnetic field, the FMR frequency, f_0, can be obtained by $f_0 = \gamma(H - 4\pi M_{s,eff})$ [24] where H is the magnitude of media switching field with an absence of MAMR and $M_{s,eff}$ is the effective saturation magnetization of the bilayer ECC-BPM. The FMR itself can exert additional energy on the magnetization through the resonance phenomenon, which causes the additional magnetization precession. The magnitude of this added energy depends on the frequency of the external microwave field. The maximum added energy occurs when the f_{ac} is synchronized with the FMR frequency of the system, which typically provides a significant reduction in H_{SW}. In addition, the reason why the media with MAMR have a lower H_{SW} than the conventional media is that the magnetization precession of the media subjected to the microwave field receives the additional energy exerted by the microwave field. The magnetization of the hard layer, therefore, requires lower H_{SW} for switching than in a case without MAMR. In particular, this additional torque energy becomes massive when the frequency of MAMR corresponds with the precession frequency of its FMR.

A comparison of the hysteresis loops of the single-layer BPM and the bilayer ECC-BPM for cases without and with 5 GHz MAMR is indicated in Figure 4. The A_{dot} of 0.14 pJ/m was selected for this characterization for both media since this value previously demonstrates the lowest H_{SW} under MAMR. In this figure, the H_{SW} is determined at the point that $M = -0.8M_s$. From the results, the single-layer BPM has the H_{SW} of 27 kOe and 26 kOe for cases without and with MAMR, respectively. By adding the Fe soft layer to the single-layer BPM, the H_{SW} of the bilayer ECC-BPM for the cases without and with MAMR was reduced to 13.3 kOe and 12.2 kOe, respectively, which are significantly lower than that of the single-layer BPM. To provide more information about the magnetization orientation of the media, the magnetic domain of four media was characterized at the points where $H_{dc} = -18$ kOe since this particular H_{dc} can indicate the difference between them. Figure 5a,d show the magnetic domain from the cross-sectional side view of media at the points shown by circles (a–d) in Figure 4, respectively. It is seen that the magnetization of the single-layer BPM with MAMR contains more reversed magnetization than that without MAMR, demonstrating the effects of MAMR assisting the magnetization switching. Additionally, the magnetization of the bilayer ECC-BPMs is greatly reversed when compared to the single-layer BPM. The use of the MAMR in the bilayer ECC-BPM indicates a slightly better reversal of magnetization.

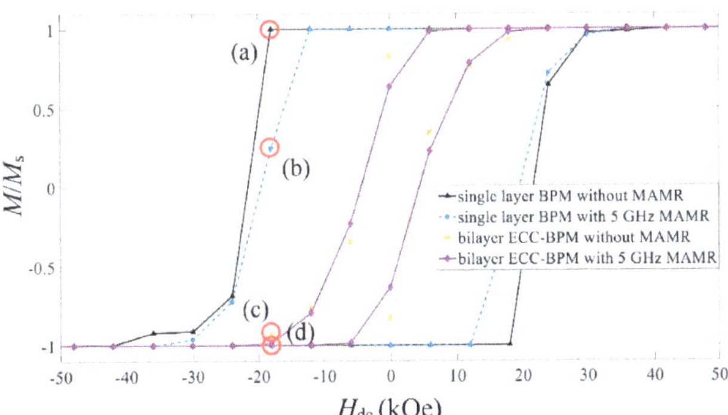

Figure 4. The hysteresis loops of the single-layer BPM and the bilayer ECC-BPM with and without MAMR ($A_{dot} = 0.14$ pJ/m for both media).

From overall evaluations, the lowest H_{SW} found in this work is 12.2 kOe, achieved by the proposed bilayer ECC-BPM with an A_{dot} of 0.14 pJ/m using a 5 GHz MAMR, which is below the maximum write head field existed in the literature. Therefore, this proposed bilayer ECC-BPM could be another choice as the magnetic media for an areal density of 4 Tb/in^2 of data storage technology. It should be noted that the suitable value of A_{dot} and f_{ac} for other media and areal densities recording systems needs to be carefully optimized. The theoretical findings can be the guidelines for its experimental verification as well as further development of magnetic media in the future.

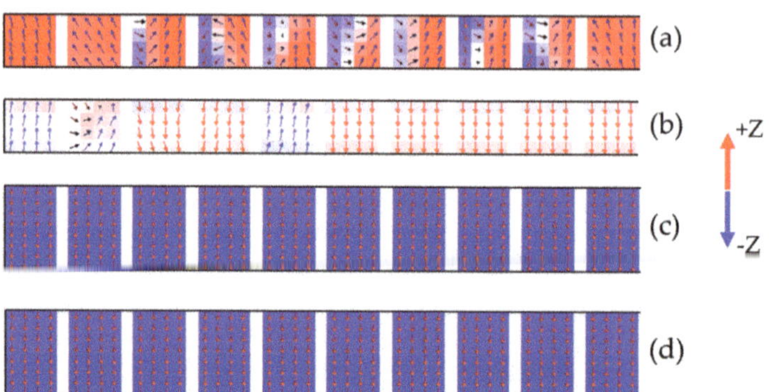

Figure 5. Magnetic domain from cross-sectional side view of (**a**) Single-layer BPM; (**b**) Single-layer BPM with MAMR; (**c**) Bilayer ECC-BPM; (**d**) Bilayer ECC-BPM with MAMR at $H_{dc} = -18$ kOe.

4. Conclusions

In this work, the $L1_0$-FePt single-layer BPM and $L1_0$-FePt/Fe bilayer ECC-BPM with MAMR technology were proposed to improve the writability of the magnetic media at a 4 Tb/in^2 recording density. It was found that the H_{sw} bilayer ECC-BPM was significantly lower than that of the single-layer BPM. The H_{sw} of those media could be increasingly reduced by using MAMR. The suitable frequency of the applied microwave field was 5 GHz, which was consistent with the FMR of the system. The suitable value of A_{dot} for the proposed bilayer ECC-BPM under MAMR was 0.14 pJ/m. The physical explanation of the H_{sw} of the medium regarding a variation of MAMR and A_{dot} was given. The hysteresis loops and magnetic domain of the medium have been considered to provide further details on the simulation results. The lowest H_{sw} found in this proposed medium is 12.2 kOe, achieved by the bilayer ECC-BPM with an A_{dot} of 0.14 pJ/m using a 5 GHz MAMR. Findings can be used for the future development of ultra-high areal densities magnetic recording technology.

Author Contributions: Conceptualization, P.K., N.W., and A.K. (Arkom Kaewrawang); methodology, P.K., N.W., and A.K, (Arkom Kaewrawang); software, N.W. C.J., and A.K. (Arkom Kaewrawang); validation, P.K., N.W., C.J., and A.K. (Arkom Kaewrawang); formal analysis, P.K., N.W., and A.K. (Arkom Kaewrawang); investigation, N.W. and A.K. (Arkom Kaewrawang); data curation, P.K., N.W., and A.K. (Arkom Kaewrawang); writing—original draft preparation, P.K. and N.W.; writing—review and editing, P.K., N.W., C.J., and A.K. (Arkom Kaewrawang); visualization, N.W., C.J., A.K. (Arkom Kaewrawang), A.S., and A.K. (Arkom Kaewrawang); supervision, A.K. (Arkom Kaewrawang) and A.S.; project administration, A.K. (Arkom Kaewrawang); funding acquisition, P.K and A.K. (Arkom Kaewrawang) All authors have read and agreed to the published version of the manuscript.

Funding: This work was financially supported by the Thailand Research Fund (TRF) and Khon Kaen University (Grant No. TRG5780125), and the Office of the Permanent Secretary, Ministry of Higher Education, Science, Research and Innovation (Grant No. RGNS 63-058).

Acknowledgments: The authors would like to thank Warunee Tipcharoen and Edanz Group Global Ltd.

Conflicts of Interest: The authors declare no conflict of interest.

References

1. Ross, C. Patterned Magnetic Recording Media. *Annu. Rev. Mater. Res.* **2001**, *31*, 203–235. [CrossRef]
2. Khunkitti, P.; Siritaratiwat, A.; Kaewrawang, A.; Mewes, T.; Mewes, C.; Kruesubthaworn, A. Electromagnetic interference-induced instability in CPP-GMR read heads. *J. Magn. Magn. Mater.* **2016**, *412*, 42–48. [CrossRef]
3. Nagasaka, K. CPP-GMR technology for magnetic read heads of future high-density recording systems. *J. Magn. Magn. Mater.* **2009**, *321*, 508–511. [CrossRef]
4. Khunkitti, P.; Pituso, K.; Chaiduangsri, N.; Siritaratiwat, A. Optimal Sizing of CPP-GMR Read Sensors for Magnetic Recording Densities of 1–4 Tb/in^2. *IEEE Access* **2021**, *9*, 130758–130766. [CrossRef]

5. Khunkitti, P.; Kaewrawang, A.; Siritaratiwat, A.; Mewes, T.; Mewes, C.K.A.; Kruesubthaworn, A. A novel technique to detect effects of electromagnetic interference by electrostatic discharge simulator to test parameters of tunneling magnetoresistive read heads. *J. Appl. Phys.* **2015**, *117*, 17A908. [CrossRef]
6. Khunkitti, P.; Siritaratiwat, A.; Pituso, K. Free Layer Thickness Dependence of the Stability in $Co_2(Mn_{0.6}Fe_{0.4})$Ge Heusler Based CPP-GMR Read Sensor for Areal Density of 1 Tb/in^2. *Micromachines* **2021**, *12*, 1010. [CrossRef] [PubMed]
7. Wang, J.-P.; Shen, W.; Bai, J. Exchange coupled composite media for perpendicular magnetic recording. *IEEE Trans. Magn.* **2005**, *41*, 3181–3186. [CrossRef]
8. Wang, Y.; Erden, M.F.; Victora, R.H. Novel system design for readback at 10 terabits per square inch user areal density. *IEEE Magn. Lett.* **2012**, *3*, 4500304.
9. Zhao, B.; Xue, H.; Zhu, Z.; Ren, Y.; Jin, Q.Y.; Zhang, Z. Interlayer modulation on the dynamic magnetic properties of L10-FePt/NM/[CoNi]5 composite film structures. *Appl. Phys. Lett.* **2019**, *115*, 062401. [CrossRef]
10. Pituso, K.; Khunkitti, P.; Tongsomporn, D.; Kruesubthaworn, A.; Chooruang, K.; Siritaratiwat, A.; Kaewrawang, A. Simulation of magnetic footprints for heat assisted magnetic recording. *Eur. Phys. J. Appl. Phys.* **2017**, *78*, 20301. [CrossRef]
11. Pituso, K.; Kaewrawang, A.; Buatong, P.; Siritaratiwat, A.; Kruesubthaworn, A. The temperature and electromagnetic field distributions of heat-assisted magnetic recording for bit-patterned media at ultrahigh areal density. *J. Appl. Phys.* **2015**, *117*, 17C501. [CrossRef]
12. Tipcharoen, W.; Kaewrawang, A.; Siritaratiwat, A. Design and Micromagnetic Simulation of Fe/L10-FePt/Fe Trilayer for Exchange Coupled Composite Bit Patterned Media at Ultrahigh Areal Density. *Adv. Mater. Sci. Eng.* **2015**, *2015*, 504628. [CrossRef]
13. Krone, P.; Makarov, D.; Schrefl, T.; Albrecht, M. Exchange coupled composite bit patterned media. *Appl. Phys. Lett.* **2010**, *97*, 82501. [CrossRef]
14. Boone, C.; Katine, J.A.; Marinero, E.E.; Pisana, S.; Terris, B.D. Microwave-Assisted Magnetic Reversal in Perpendicular Media. *IEEE Magn. Lett.* **2012**, *3*, 3500104. [CrossRef]
15. Fal, T.J.; Camley, R.E. Microwave assisted switching In bit patterned media: Accessing multiple states. *Appl. Phys. Lett.* **2010**, *97*, 122506. [CrossRef]
16. Bai, X.; Zhu, J.-G. Effective Field Analysis of Segmented Media for Microwave-Assisted Magnetic Recording. *IEEE Magn. Lett.* **2017**, *8*, 1–4. [CrossRef]
17. Greaves, S.J.; Chan, K.S.; Kanai, Y. Areal Density Capability of Dual-Structure Media for Microwave-Assisted Magnetic Re-cording. *IEEE Trans. Magn.* **2019**, *55*, 6701509. [CrossRef]
18. Tanaka, T.; Kurihara, K.; Ya, X.; Kanai, Y.; Bai, X.; Matsuyama, K. MAMR writability and signal-recording characteristics on granular exchange-coupled composite media. *J. Magn. Magn. Mater.* **2021**, *529*, 167884. [CrossRef]
19. Shiroishi, Y.; Fukuda, K.; Tagawa, I.; Iwasaki, H.; Takenoiri, S.; Tanaka, H.; Mutoh, H.; Yoshikawa, N. Future Options for HDD Storage. *IEEE Trans. Magn.* **2009**, *45*, 3816–3822. [CrossRef]
20. Donahue, M.J.; Porter, D.G. OOMMF User's Guide, Release 1.2a3. Available online: https://math.nist.gov/oommf/ftp-archive/doc/userguide12a3_20021030.pdf (accessed on 15 October 2021).
21. Barmak, K.; Coffey, K.R.; Thiele, J.-U.; Kim, J.; Lewis, L.H.; Toney, M.F.; Kellock, A.J. Stoichiometry-anisotropy connections in epitaxial L10 FePt(001) films. *J. Appl. Phys.* **2004**, *95*, 7501–7503. [CrossRef]
22. Wang, F.; Xu, X.-H.; Liang, Y.; Zhang, J.; Zhang, J. Perpendicular L10-FePt/Fe and L10-FePt/Ru/Fe graded media obtained by post-annealing. *Mater. Chem. Phys.* **2011**, *126*, 843–846. [CrossRef]
23. Laval, M.; Bonnefois, J.J.; Bobo, J.F.; Issac, F.; Boust, F. Microwave-assisted switching of NiFe magnetic microstructures. *J. Appl. Phys.* **2009**, *105*, 73912. [CrossRef]
24. Kittel, C. *Introduction to Solid State Physics*; John Wiley & Sons: Hoboken, NJ, USA, 2011.

micromachines

MDPI

Article

Optimization of a Spin-Orbit Torque Switching Scheme Based on Micromagnetic Simulations and Reinforcement Learning

Roberto L. de Orio [1,*], Johannes Ender [2], Simone Fiorentini [2], Wolfgang Goes [3], Siegfried Selberherr [1] and Viktor Sverdlov [2]

1 Institute for Microelectronics, TU Wien, Gußhausstraße 27-29/E360, 1040 Vienna, Austria; Selberherr@TUWien.ac.at

2 Christian Doppler Laboratory for Nonvolatile Magnetoresistive Memory and Logic at the Institute for Microelectronics, TU Wien, 1040 Vienna, Austria; ender@iue.tuwien.ac.at (J.E.); fiorentini@iue.tuwien.ac.at (S.F.); sverdlov@iue.tuwien.ac.at (V.S.)

3 Silvaco Europe Ltd., Cambridge PE27 5JL, UK; wolfgang.goes@silvaco.com

* Correspondence: orio@iue.tuwien.ac.at

Abstract: Spin-orbit torque memory is a suitable candidate for next generation nonvolatile magnetoresistive random access memory. It combines high-speed operation with excellent endurance, being particularly promising for application in caches. In this work, a two-current pulse magnetic field-free spin-orbit torque switching scheme is combined with reinforcement learning in order to determine current pulse parameters leading to the fastest magnetization switching for the scheme. Based on micromagnetic simulations, it is shown that the switching probability strongly depends on the configuration of the current pulses for cell operation with sub-nanosecond timing. We demonstrate that the implemented reinforcement learning setup is able to determine an optimal pulse configuration to achieve a switching time in the order of 150 ps, which is 50% shorter than the time obtained with non-optimized pulse parameters. Reinforcement learning is a promising tool to automate and further optimize the switching characteristics of the two-pulse scheme. An analysis of the impact of material parameter variations has shown that deterministic switching can be ensured for all cells within the variation space, provided that the current densities of the applied pulses are properly adjusted.

Keywords: spin-orbit torque MRAM; reinforcement learning; two-pulse switching scheme; magnetic field-free switching; machine learning

Citation: de Orio, R.L.; Ender, J.; Fiorentini, S.; Goes, W.; Selberherr, S.; Sverdlov, V. Optimization of a Spin-Orbit Torque Switching Scheme Based on Micromagnetic Simulations and Reinforcement Learning. *Micromachines* **2021**, *12*, 443. https://doi.org/10.3390/mi12040443

Academic Editor: Sumeet Kumar Gupta

Received: 12 March 2021
Accepted: 13 April 2021
Published: 15 April 2021

1. Introduction

Spin-transfer torque magnetoresistive random access memory (STT-MRAM) is currently the state-of-the-art MRAM technology, entering volume production at all major foundries [1–6]. It is an emerging nonvolatile technology suitable for future universal memory applications. One of its key advantages is that it is compatible with CMOS technology, so it can be straightforwardly embedded in circuits [7]. It is promising not only for standalone, but also for embedded memory applications as replacement of conventional volatile CMOS-based and nonvolatile flash memories in systems on chip. STT-MRAM can be integrated in a broad range of applications, from Internet-of-Things to automotive applications [3] and last level caches [8–10]. Recently, 1Gb standalone [11] and embedded STT-MRAM solutions [2,4,12,13] have been reported and STT-MRAM operation with a timing of a few nanoseconds has been demonstrated [8]. However, in order to further reduce the timing below the nanosecond range, the required current density becomes quite large. This creates an important limitation, since large currents flowing through the thin tunnel oxide of a magnetic tunnel junction (MTJ) lead to reliability issues, reducing the MRAM endurance.

Spin-orbit torque (SOT) MRAM is a promising nonvolatile memory candidate outperforming STT-MRAM for ultra-fast operation [14]. In SOT-MRAM, the large current required for the writing operation does not flow through the MTJ. Switching is accomplished by applying a current through a heavy metal wire attached to the magnetic free layer (FL). Thus, it can operate with a sub-nanosecond timing retaining excellent endurance [15–17]. These properties make SOT-MRAM particularly interesting for nonvolatile replacement of the classical static random access memory (SRAM) used in caches. It should be pointed out, however, that deterministic SOT switching of a perpendicularly magnetized FL requires an external magnetic field [18]. Several field-free schemes have been proposed to circumvent this issue, usually at the cost of a more complex cell stack fabrication [15,16,19–23].

We consider an alternative field-free scheme, in which a purely electrical control of the switching process is realized by applying current pulses to two orthogonal heavy metal wires [24]. A proper configuration of the current pulses is able to reverse the perpendicularly magnetized FL [25]. Nevertheless, an interesting question arises: how can one determine a pulse sequence that leads to optimal switching? Searching for such a pulse sequence requires a very large number of experiments and/or simulations. Ideally, this task can be outsourced to an algorithm and performed in a guided and automated way.

Machine learning (ML) has been increasingly applied to the solution of physics-based problems [26] and has already been used to solve fundamental micromagnetic problems, such as the computation of the magnetization dynamics of a thin film [27] and of the magnetic microstructure of a single magnetic body [28]. Recently, an ML model was applied to identify the regime of field-free SOT switching as a function of the magnitude of the applied current density, the nanomagnet size, and the interfacial Dyzaloshinskii-Moriya interaction [29]. The two most used ML approaches are supervised and unsupervised learning. In turn, another sub-branch of ML so-called reinforcement learning (RL) has gained interest [30,31]. Big RL breakthroughs were achieved lately using games like chess or Go [32], but this type of learning algorithm has also been successfully applied for physics-based problems.

The RL principle of operation is based on an agent and an environment, in such a way that the agent interacts with the environment and learns how to act or take decisions to achieve a specific state or goal [30]. In other words, the agent interacts with the environment by performing actions that cause the environment to move from one state to another. Once the environment moves to a new state, it informs about its state and returns a reward to the agent. Based on this information, the agent can decide to take actions to maximize the cumulative reward received over time. During this process, the agent learns how to achieve the given objective.

In this work, we combine the two-current pulse switching scheme with an RL algorithm to optimize the switching of a spin-orbit torque memory cell. We demonstrate that the reinforcement learning implementation can find an optimal sequence and timing for the current pulses in order to achieve faster switching in comparison to a conventional combination of pulse parameters.

2. Spin-Orbit Torque Memory Cell and Switching Dynamics

The two-pulse switching scheme for a SOT memory cell is depicted in Figure 1. The cell is formed by growing a perpendicularly magnetized FL on top of a heavy metal wire (NM1), where a first current pulse is applied to generate the initial SOT on the FL. On the right part of the cell, a second, orthogonal heavy metal wire (NM2) is placed on top of the FL, and a second current pulse is applied through it. The SOT generated due to this second pulse acts on the FL to complete the magnetization switching of the memory cell [33]. The NM1/FL/NM2 stack composes the structural part used for the writing operation of the memory cell. In the left part of the cell, next to the SOT writing stack, an MTJ is grown on top of the FL, which is required for carrying out the reading operation of the memory cell via measurement of the tunneling magnetoresistance.

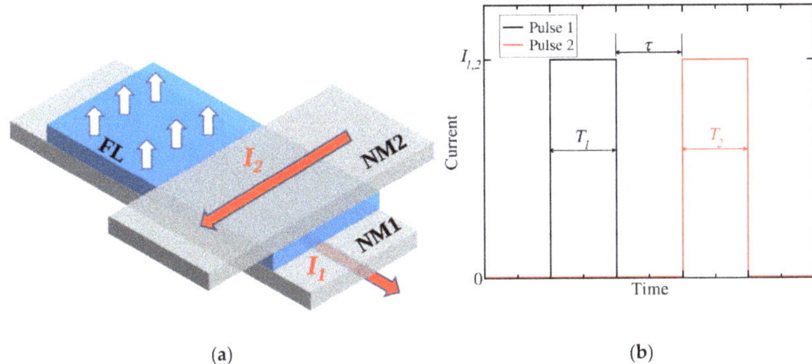

(a) (b)

Figure 1. (**a**) SOT-MRAM cell for switching based on (**b**) two orthogonal current pulses. Pulse 1 is applied to the NM1 wire and Pulse 2 is applied to the NM2 wire. $I_{1,2}$ is the current amplitude and $T_{1,2}$ is the width of the first/second pulse. τ represents the delay or overlap between the pulses.

In order to carry out micromagnetic simulations of the two-pulse SOT switching scheme, the magnetization dynamics is described by the Landau-Lifshitz-Gilbert (LLG) equation

$$\begin{aligned}\frac{\partial \mathbf{m}}{\partial t} &= -\gamma\mu_0\, \mathbf{m}\times\mathbf{H_{eff}} + \alpha\mathbf{m}\times\frac{\partial \mathbf{m}}{\partial t}\\ &-\gamma\frac{\hbar}{2e}\frac{\theta_{SH}j_1}{M_S d}[\mathbf{m}\times(\mathbf{m}\times\mathbf{y})]\Theta(t,T_1)\\ &+\gamma\frac{\hbar}{2e}\frac{\theta_{SH}j_2}{M_S d}[\mathbf{m}\times(\mathbf{m}\times\mathbf{x})]\Theta(t,T_1,T_2)\end{aligned} \quad (1)$$

where $\mathbf{m}$ is the normalized magnetization, γ is the gyromagnetic ratio, μ_0 is the vacuum permeability, α is the Gilbert damping factor, and M_S is the saturation magnetization. $\mathbf{H_{eff}}$ is an effective magnetic field, which includes the exchange field, the uniaxial perpendicular anisotropy field, the demagnetization field, the current-induced field, and the stochastic thermal field at 300 K. The last two terms on the right-hand side of the LLG equation describe the SOT generated by the applied current pulses through the NM1 and the NM2 wire, respectively, where e is the elementary charge, $\hbar$ is the reduced Plank constant, θ_{SH} is an effective Hall angle, $j_{1,2}$ is the current density of the first/second pulse, d is the FL thickness, and $\Theta(\cdot)$ is a function which determines when each pulse is active.

Equation (1) is solved numerically using a micromagnetic simulation software developed in-house [34] based on the finite difference method. The simulation parameters are given in Table 1.

Table 1. Simulation parameters. Heavy metal wires of b-tungsten and a magnetic FL of CoFeB on MgO are assumed [18].

Parameter	Value
Saturation magnetization, M_S	1.1×10^6 A/m
Exchange constant, A	1.0×10^{-11} J/m
Perpendicular anisotropy, K	8.4×10^5 J/m^3
Gilbert damping factor, α	0.035
Spin Hall angle, θ_{SH}	0.3
Thermal stability factor, Δ	45
Free layer dimensions	40 nm $\times$ 20 nm $\times$ 1.2 nm
NM1: $w_1\times l$	20 nm $\times$ 3 nm
NM2: $w_2\times l$	20 nm $\times$ 3 nm

3. Reinforcement Learning for the Two-Pulse Spin-Orbit Torque Switching

Figure 2 shows the RL setup implemented for performing the learning experiments with the two-pulse switching scheme. The environment consists of our in-house tool, which provides the simulation of the memory cell switching and returns the current state of the simulation together with a reward after every iteration. The used deep Q-network (DQN) algorithm [31] incorporates a neural network to approximate a function for mapping states to actions. An existing Python library providing the RL capabilities has been employed [35]. Here, the goal of our RL implementation is to determine the pulse configuration which results in the shortest switching time, defined as the time when the perpendicular component of the magnetization vector reaches −0.5, i.e., $m_z = -0.5$.

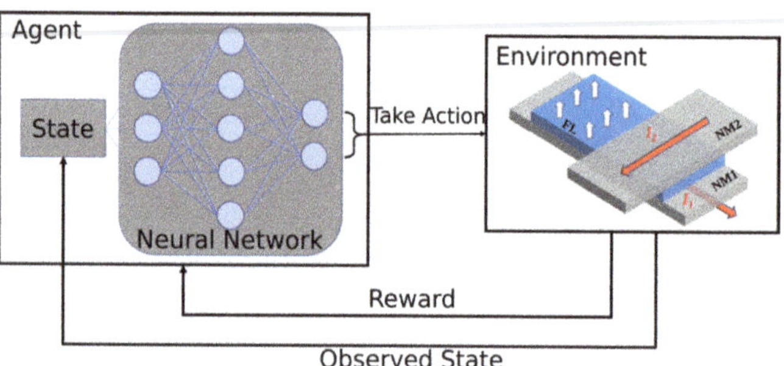

Figure 2. Reinforcement learning setup for the two-pulse switching scheme. The micromagnetic simulation of the memory cell provides the environment, with which the agent interacts and takes actions to achieve the fastest magnetization switching.

The state vector returned from the environment after every iteration consists of 11 variables: the average of the three magnetization vector components (m_x, m_y, m_z), the difference of each component to the previous iteration (Δm_x, Δm_y, Δm_z), the average component of the effective magnetic field ($H_{eff,x}$, $H_{eff,y}$, $H_{eff,z}$), and two variables indicating whether the first and the second pulse are active or not. Based on the state information, the learning agent deduces which action to take. It is important that the dynamics of the magnetization vector, given by (Δm_x, Δm_y, Δm_z), is taken into account, so the direction in which the magnetization is moving is known. In this way, the agent can decide on the best action to take to drive the switching as fast as possible. Our setup allows the agent to take four different actions, namely, setting both pulses off, setting both pulses on, turn the first pulse on with the second off, or turn the first pulse off and the second pulse on. If a pulse is on, it means that current has been applied to the corresponding heavy metal wire and a spin torque is applied to the magnetization of the FL.

The rewarding scheme is critical for the RL approach, because it is the main factor which leads the learning algorithm in the right direction and the agent to select the best actions to achieve the target. The reward is an integer value returned by the environment, indicating whether the actions performed by the agent were good or bad. For the SOT switching, the rewarding scheme is chosen such that a shorter switching time corresponds to a higher reward, since the RL algorithm tries to maximize the cumulative reward during the learning process. Here, a reward of −1 is given for every simulation step in which the target, $m_z = -0.5$, has not been reached yet. We define $t_{max} = 1$ ns as an upper limit for the simulation time. If the target is not reached within this time, the learning episode is terminated and a new one is started. On the other hand, if the target is reached at a time t_{final} before t_{max}, a positive reward of $(t_{max} - t_{final})/\Delta t$ is returned, where Δt is the simulation time-step. In this way, the rewarding scheme is a complementary measure of the number of time-steps required to reach the switching. The smaller the number of time-steps needed to switch, the shorter the switching time is and, therefore, the larger the reward is.

4. Results and Discussion

4.1. Numerical Simulations

Micromagnetic simulations of the switching dynamics of the two-pulse SOT scheme, as described in Section 2, were carried out. We start by investigating the impact of the pulse configuration on the magnetization dynamics. In particular, the current densities of the first and the second current pulse are fixed at $j_1 = 2.7 \times 10^{12}$ A/m^2 and $j_2 = 1.3 \times 10^{12}$ A/m^2, respectively, while the pulse durations T_1 and T_2 can be modified (c.f. Figure 1). A perfect synchronization between the pulses is considered, i.e., the second current pulse is turned on immediately after the first pulse is turned off. Thus, there is no delay or overlap between the pulses ($\tau = 0$). This constraint will be lifted in Section 4.2, where the results of the RL approach are discussed.

Figure 3 shows the perpendicular component of the magnetization (m_z) as a function of time for different widths of the first current pulse, while the second pulse width is kept fixed at $T_2 = 100$ ps. In order to account for the thermal spread resulting from the stochastic thermal field at room temperature, a total of 50 realizations are considered for each simulation condition. The curves shown in Figure 3 represent the average of these 50 realizations. One can clearly see that, depending on the width of the first pulse, the magnetization dynamics changes significantly, and so does the switching behavior. Here, the pulse sequence and the timing lead to successful magnetization reversal, when the width of the first pulse is short, while switching does not occur for larger values of T_1.

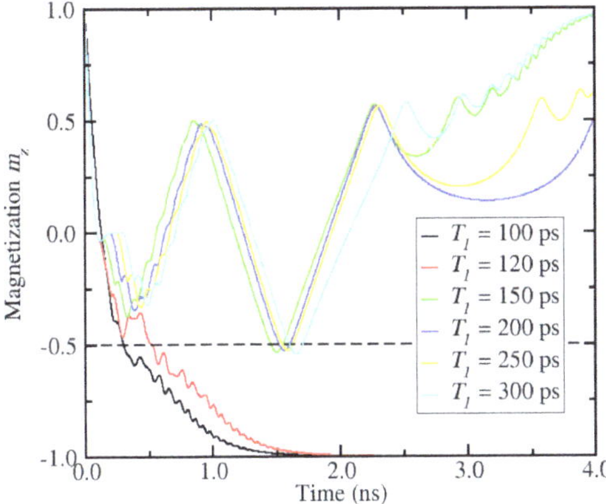

Figure 3. Perpendicular component of the magnetization vector (average of 50 realizations) as a function of time for various durations of the first pulse, T_1. The simulation parameters are found in Table 1 and $j_1 = 2.7 \times 10^{12}$ A/m^2, $j_2 = 1.3 \times 10^{12}$ A/m^2, and $T_2 = 100$ ps. j_1 and T_1 are the current density and the duration of the first pulse, respectively, and j_2 and T_2 are the current density and the duration of the second pulse (c.f. Figure 1b). The dashed line represents the switching threshold.

Next, we reverse the analysis and fix the first current pulse width at $T_1 = 150$ ps, while the width of the second current pulse is varied. The resulting magnetization dynamics is shown in Figure 4. As in the previous results, switching is obtained depending on the value of T_2. In contrast to the previous scenario, successful switching is observed as the second pulse width becomes longer.

The above results suggest that the configuration of the pulse sequence has an important impact on the switching characteristics of the cell, in such a way that variations of the pulse configuration can lead to either switching or non-switching schemes. To further understand this impact, we performed simulations for various combinations of pulses and evaluate the switching probability. The results are shown in Figure 5, which plots the switching probability as a function of the first and the second pulse width. In general, for short values of T_2 ($\leq$150 ps), the switching probability depends largely on the first pulse width, i.e., it depends on the particular pulse sequence and small changes of the pulses can yield successful or non-successful magnetization switching. In turn, increasing T_2 beyond ~200 ps, the switching probability tends to 1, becoming practically insensitive to the duration of the first pulse.

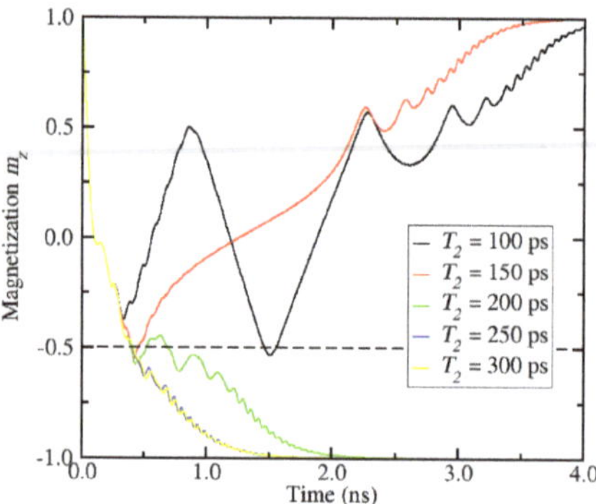

Figure 4. Switching dynamics for different values of the second pulse duration T_2 for a fixed first pulse width T_1 = 150 ps. The dashed line represents the switching threshold.

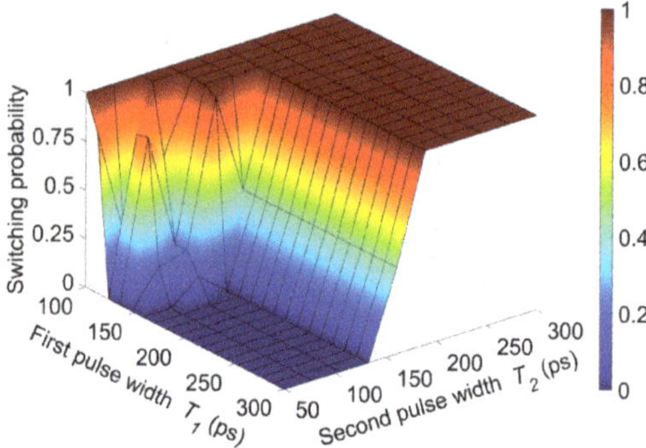

Figure 5. Switching probability as a function of the first and the second current pulse widths, T_1, T_2. For short pulse widths, precise pulse schemes are required to obtain deterministic switching.

From the previous analysis, we are able to determine pulse parameters that lead to deterministic switching of the memory cell. However, this does not guarantee that these parameters produce fast switching. Now we would like to find the pulse sequence which leads to the fastest possible switching. In order to accomplish that, we have to evaluate many more combinations of pulse sequences than those considered before. It should be pointed out that the previous results were obtained by manually running a total of 180 micromagnetic simulations. Considering that 50 realizations (due to the stochastic thermal field) are carried out for each pulse sequence combination, the number of switching simulations increases to 9000, even though delays or overlaps between the pulses are still not considered. Thus, taking into account all possible variations of pulse parameters results in an exponential increase of the required number of simulations, which makes a manual optimization of the switching intractable. Here, the RL setup described in Section 3 is extremely useful, offering a powerful methodology for searching the fastest switching condition in a guided way.

4.2. Reinforcement Learning Experiments

RL is applied with the goal of achieving the fastest magnetization switching, namely to achieve the shortest switching time, which is determined by the time when the condition $m_z = -0.5$ is reached. The agent searches for a pulse sequence and combination of the first and the second pulse duration, T_1, T_2, which lead to the shortest switching time. The actions performed by the agent (c.f. Figure 2) have been restricted to facilitate the learning process, thus it can switch on and off each pulse individually. However, the pulse synchronization constraint of the previous section is now relaxed, so that the current pulses are allowed to overlap or be delayed. The minimum pulse width is limited to 100 ps and the amplitude of the pulse is fixed to 130 μA and 100 μA for the first and the second current pulse, respectively. A learning episode is finished once $m_z = -0.5$ or the time has reached 1 ns.

The results of the learning process of our RL setting are shown in Figures 6 and 7, respectively. Figure 6 reports the switching time over the course of the learning period for 20 independent learning runs, where each run encompasses 10^6 learning steps. During an initial exploration phase, the action selection by the agent is not greedy, i.e., an action is not selected with the purpose of accumulating the highest reward, but the agent takes a random action to explore the state-action space. Furthermore, different random seeds are used for initializing the neural network weights. A general trend can, however, be observed, which is the reduction of the switching time as the number of learning steps increases. Initially, the switching times are distributed around 400–500 ps, but as the number of learning steps increases, several runs reach switching times in the 200–300 ps range.

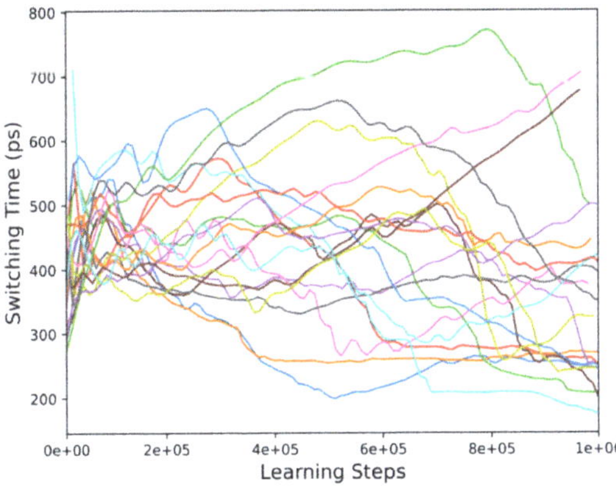

Figure 6. Switching time over the course of the learning period of 20 independent runs. As the number of learning steps increases, there is a trend towards switching time reduction.

The switching time decrease with the learning progress can be better visualized in Figure 7, which shows the mean switching time and the reward as a function of the number of learning steps of the six best learning runs. First, an increase of the switching time is observed, which is a consequence of the initial focus on exploration of the state-action space previously mentioned. Then, over the course of 10^6 learning steps, the mean switching time reduces to around 240 ps. The direct relationship between the switching time and the accumulated reward is readily demonstrated in Figure 7. As the switching time decreases, the accumulated reward increases, indicating that the agent has learned a better policy to select actions which can switch the memory cell faster. It should be pointed out that single runs were able to achieve an even better policy, which resulted in a minimum switching time of about 146 ps.

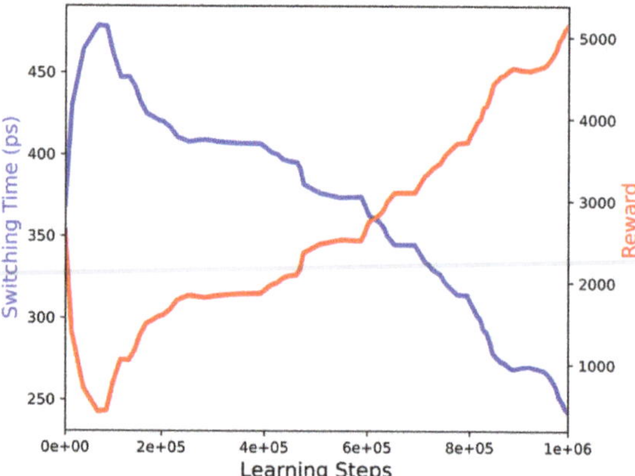

Figure 7. Learning curve showing the mean switching time and reward over 10^6 time steps. It shows that, the faster the cell switches, the larger is the accumulated reward during learning.

The pulse configuration learned by the DQN algorithm and the resulting magnetization dynamics are shown in Figure 8. The current pulses through the NM1 and the NM2 wire are turned on simultaneously right in the beginning of the simulation. After 100 ps, the first pulse is turned off and the magnetization component m_z drops below the -0.5 threshold. Once this threshold is achieved, no further action is taken and the second current pulse is kept active for the rest of the simulation. This generates a SOT which acts on the FL under the NM2 wire, resulting in an average perpendicular magnetization component of about -0.8. Thus, the magnetization of the FL is not fully reversed to -1. This demonstrates the importance of the rewarding scheme and the general setup of the RL experiment. As the RL agent was rewarded for finishing the learning episode as fast as possible and the episode was considered finished as soon as the -0.5 threshold was reached, the agent learned how to achieve the threshold and did not take any action afterwards.

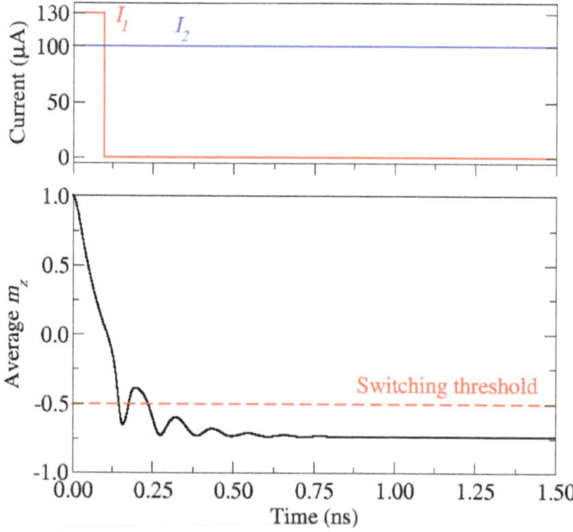

Figure 8. Pulse sequence learned by the DQN agent. I_1 is the current amplitude of the first pulse applied to the NM1 wire and I_2 is the current amplitude of the second pulse applied to the NM2 wire.

Figure 9 shows the dynamics of the magnetization component m_z considering different variations of the learned pulse configuration. In the learned model, the second pulse is now switched off after $m_z = -0.5$ is reached, which guarantees that the magnetization reversal is completed. The variations consisted of extending the first pulse and/or delaying the second pulse. A comparison of the magnetization dynamics with the learned model is given in Figure 9. One can observe that the learned configuration (black curve) leads indeed to the fastest switching. In turn, in the scenario with a longer first pulse, for which the pulses are almost perfectly overlapping, switching does not occur (red curve). The modified pulse sequences, represented by the green and blue curves, also lead to switching of the cell, however with longer switching times.

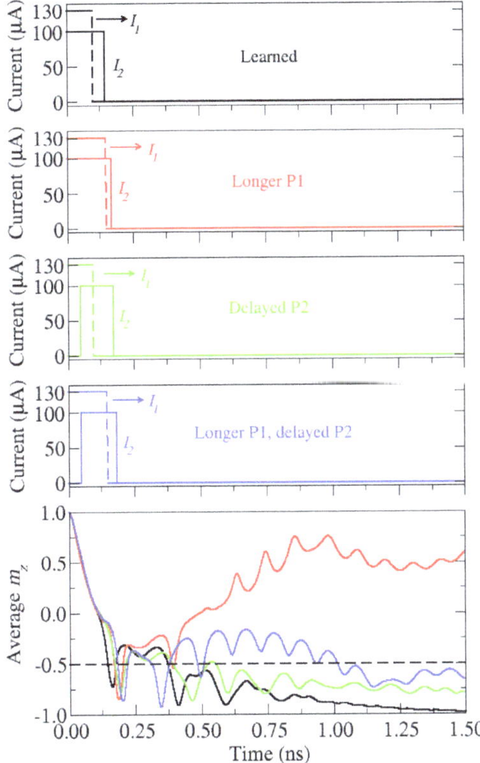

Figure 9. Comparison of different pulse configurations. The learned model is compared with modified ones. The learned pulse configuration leads to the fastest switching. I_1 is the current amplitude of the first pulse, P1, applied to the NM1 wire and I_2 is the current amplitude of the second pulse, P2, applied to the NM2 wire. The dashed line represents the switching threshold.

The robustness of the switching for the learned scheme is confirmed in Figure 10, for which 50 realizations under influence of the stochastic thermal field are reported. The variations between the different realizations are small and all of them switch, which shows that the learned scheme results in reliable and deterministic switching. It should be pointed out that, while the RL approach was able to find a scheme for which the switching time is 146 ps, the minimum switching time obtained from the previous manual configuration of the pulse was around 300 ps. This demonstrates the potential of the RL tool in combination with micromagnetic simulation for optimizing the two-pulse SOT switching scheme.

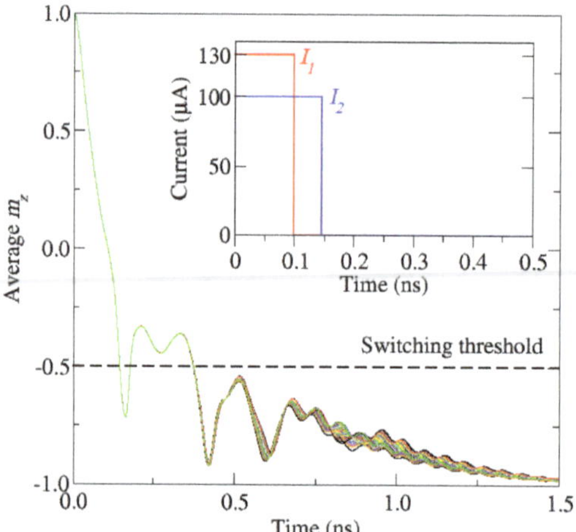

Figure 10. z-component of the magnetization of 50 switching realizations using the switching scheme found by the RL algorithm shown in the inset. I_1 is the current amplitude of the first pulse and I_2 is the current amplitude of the second pulse.

4.3. Impact of Parameter Variations

Although the fastest switching condition has been determined, variations of the pulse timing and/or of the process and material parameters of the magnetic FL can lead to slower or even non-deterministic switching. Thus, we now consider the impact of variations of the saturation magnetization and the anisotropy energy on the switching scheme.

Figure 11 shows the x, y, and z components of the magnetization vector as a function of time for $K = 8.8 \times 10^5$ J/m^3 and $M_S = 1.05 \times 10^6$ A/m, which represent a variation of 5% in relation to the nominal parameter values. In this case the cell does not switch and, more importantly, one can observe that the perpendicular component of the magnetization (m_z) does not reduce below 0.7. This means that the SOT generated by the applied current density of the first pulse ($j_1 = 2.7 \times 10^{12}$ A/m^2) is too weak to trigger the magnetization reversal. This can be explained by the fact that the variation of material parameters can change the critical current density for SOT switching. The above parameters lead to an increase of the critical current density, so that it becomes larger than the applied one. Thus, in order to switch this particular cell, the applied current has to be increased.

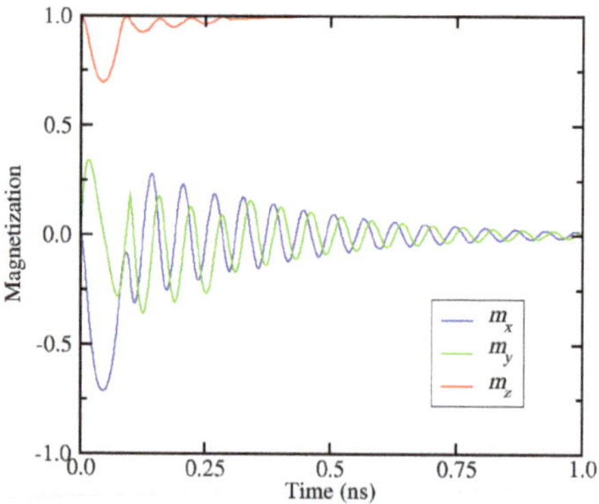

Figure 11. Magnetization dynamics for a cell with 5% variation of the perpendicular anisotropy energy and saturation magnetization in relation to the nominal values. The current density of the first pulse (2.7×10^{12} A/m^2) is smaller than the critical current density, so the cell does not switch.

Considering that different material parameter variations happen concurrently, one should expect that different cells of the same design undergoing the same fabrication process can require different current densities to trigger switching. Figure 12 shows the required current density for the first pulse to guarantee deterministic switching for various combinations of saturation magnetization and anisotropy energy. For 10% variation of the parameters, the minimum switching current density varies from 1.0×10^{12} A/m^2 to about 3.0×10^{12} A/m^2. These results indicate that, in order to switch all cells within the parameter spread, a current density of at least 3.0×10^{12} A/m^2 has to be applied for the first pulse.

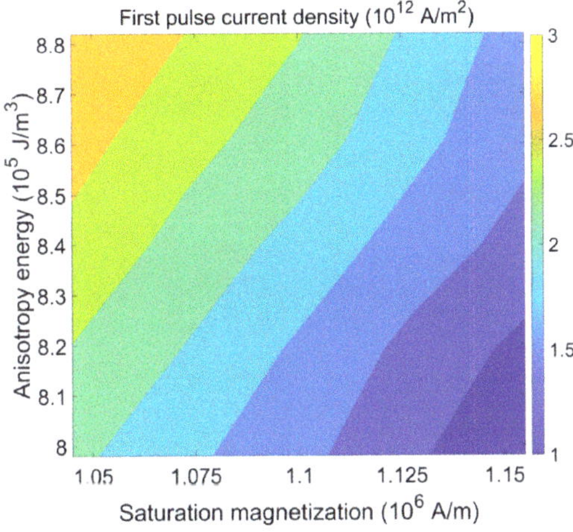

Figure 12. Minimum current density required for the first pulse to trigger reliable magnetization switching for cells with different combinations of saturation magnetization and anisotropy energy. The lowest current density is 1.0×10^{12} A/m^2 (lower right corner) and the highest value is about 3.0×10^{12} A/m^2 (upper left corner). The current density of the second pulse is 1.3×10^{12} A/m^2 and both pulse durations are set to 200 ps.

Next, the required current density for the second pulse is determined, as shown in Figure 13. We consider four combinations of anisotropy energy and saturation magnetization, denominated C1 to C4, which cover the variation space of Figure 12: $K = 8.8 \times 10^5$ J/m^3, $M_S = 1.05 \times 10^6$ A/m (C1), $K = 8.8 \times 10^5$ J/m^3, $M_S = 1.16 \times 10^6$ A/m (C2), $K = 8.0 \times 10^5$ J/m^3, $M_S = 1.05 \times 10^6$ A/m (C3), and $K = 8.0 \times 10^5$ J/m^3, $M_S = 1.16 \times 10^6$ A/m (C4), where C1 and C4 correspond to the two extreme cases, upper left and lower right corner, respectively, of Figure 12. In order to ensure 100% switching, the minimum current density required for the second pulse is about 1.0×10^{12} A/m^2.

The above analysis has allowed us to determine the minimum settings which guarantee 100% switching in the presence of cell-to-cell variations. Applying $j_1 = 3.0 \times 10^{12}$ A/m^2 and $j_1 = 1.3 \times 10^{12}$ A/m^2, the average switching realizations from parallel to anti-parallel (P-AP) as well as from anti-parallel to parallel (AP-P) configuration are reported in Figure 14, for the parameter combinations C1 to C4 and the nominal (Nom.) case, $K = 8.4 \times 10^5$ J/m^3, $M_S = 1.1 \times 10^6$ A/m. It should be pointed out that 50 realizations have been tested for each combination and all of them resulted in successful switching.

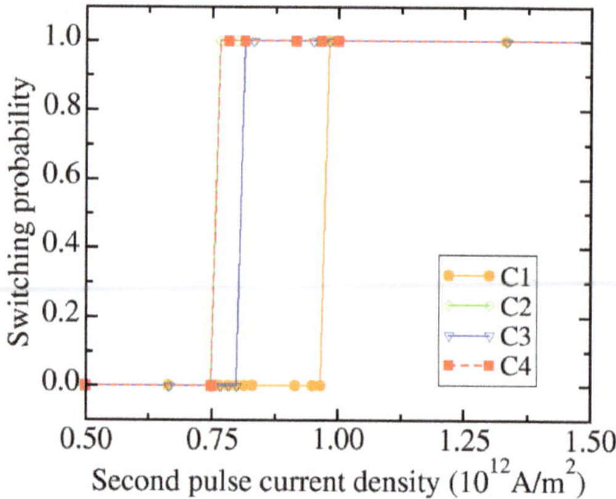

Figure 13. Switching probability as a function of the current density for the second current pulse, for four combinations of anisotropy energy and saturation magnetization. C1: $K = 8.8 \times 10^5$ J/m^3, $M_S = 1.05 \times 10^6$ A/m; C2: $K = 8.8 \times 10^5$ J/m^3, $M_S = 1.16 \times 10^6$ A/m; C3: $K = 8.0 \times 10^5$ J/m^3, $M_S = 1.05 \times 10^6$ A/m; C4: $K = 8.0 \times 10^5$ J/m^3, $M_S = 1.16 \times 10^6$ A/m. The current density of the first pulse is set to 3.0×10^{12} A/m^2 and both pulse durations are 200 ps.

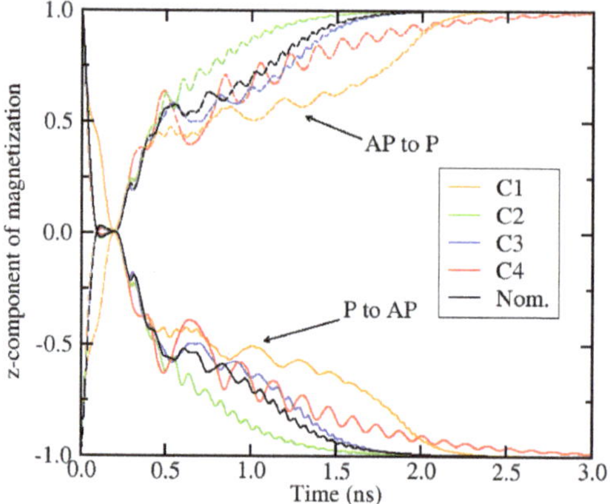

Figure 14. Average switching realizations from parallel to anti-parallel (P-AP) and anti-parallel to parallel (AP-P) for various combinations of anisotropy energy and saturation magnetization. C1: $K = 8.8 \times 10^5$ J/m^3, $M_S = 1.05 \times 10^6$ A/m; C2: $K = 8.8 \times 10^5$ J/m^3, $M_S = 1.16 \times 10^6$ A/m; C3: $K = 8.0 \times 10^5$ J/m^3, $M_S = 1.05 \times 10^6$ A/m; C4: $K = 8.0 \times 10^5$ J/m^3, $M_S = 1.16 \times 10^6$ A/m; Nom.: $K = 8.4 \times 10^5$ J/m^3, $M_S = 1.1 \times 10^6$ A/m. Each curve represents the average of 50 realizations, all of them resulting in successful switching.

5. Conclusions

We developed a reinforcement learning approach in combination with micromagnetic simulations to optimize the switching of a spin-orbit torque memory cell. The magnetization switching is accomplished with a two-current pulse scheme and it is shown that, for sub-nanosecond operation, the switching probability strongly depends on the parameters of the applied current pulses. We demonstrated that the reinforcement learning setup can determine optimal sequence and timing parameters for the current pulses, which results in the fastest switching of the memory cell. This optimal pulse sequence yielded a switching time as short as 146 ps, remarkably shorter in comparison to a switching time of 300 ps for the manually configured pulse sequence. Based on our results, reinforcement learning is a promising tool to automate and further optimize spin-orbit torque switching based on the two-pulse scheme. We analyzed the impact of material parameter variations and showed that reliable switching can be guaranteed in the presence of cell-to-cell variations, provided that the current amplitude of the pulses is adjusted.

Author Contributions: Conceptualization, R.L.d.O. and J.E.; Data curation, R.L.d.O. and J.E.; Formal analysis, R.L.d.O. and J.E.; Funding acquisition, S.S. and V.S.; Investigation, R.L.d.O. and J.E.; Methodology, R.L.d.O. and J.E.; Project administration, V.S.; Software, R.L.d.O. and J.E.; Supervision, S.S. and V.S.; Visualization, R.L.d.O. and J.E.; Writing—original draft, R.L.d.O.; Writing—review & editing, J.E., S.F., W.G., S.S., and V.S. All authors have read and agreed to the published version of the manuscript.

Funding: This research was funded by the Christian Doppler Forschungsgesellschaft, grant number 1558669. The APC was funded by the TU Wien Library through its Open Access Funding Program.

Acknowledgments: The financial support by the Austrian Federal Ministry for Digital and Economic Affairs and the National Foundation for Research, Technology and Development is gratefully acknowledged. The authors acknowledge TU Wien Library for financial support through its Open Access Funding Program.

Conflicts of Interest: The authors declare no conflict of interest.

References

1. Jew, T. MRAM in Microcontroller and Microprocessor Product Applications. In Proceedings of the 2020 IEEE International Electron Devices Meeting (IEDM), San Francisco, CA, USA, 12–18 December 2020; pp. 11.1.1–11.1.4.
2. Han, S.; Lee, J.; Shin, H.; Lee, J.; Suh, K.; Nam, K.; Kwon, B.; Cho, M.; Lee, J.; Jeong, J.; et al. 28-nm 0.08mm^2/Mb Embedded MRAM for Frame Buffer Memory. In Proceedings of the 2020 IEEE Electron Devices Meeting (IEDM), San Francisco, CA, USA, 12–18 December 2020; pp. 11.2.1–11.2.4.
3. Naik, V.B.; Yamane, K.; Lee, T.Y.; Kwon, J.H.; Chao, R.; Lim, J.; Chung, N.L.; Behin-Aein, B.; Hau, L.Y.; Zeng, D.; et al. JEDEC-Qualified Highly Reliable 22nm FD-SOI Embedded MRAM For Low-Power Industrial-Grade, and Extended Performance Towards Automotive-Grade-1 Applications. In Proceedings of the 2020 IEEE Electron Devices Meeting (IEDM), San Francisco, CA, USA, 12–18 December 2020; pp. 11.3.1–11.3.4.
4. Shih, Y.C.; Lee, C.F.; Chang, Y.A.; Lee, P.H.; Lin, H.J.; Chen, Y.L.; Lo, C.-P.; Lin, K.F.; Chiang, T.W.; Lee, Y.J.; et al. A Reflow-Capable, Embedded 8Mb STT-MRAM Macro with 9ns Read Access Time in 16nm FinFet Logic CMOS Process. In Proceedings of the 2020 IEEE Electron Devices Meeting (IEDM), San Francisco, CA, USA, 12–18 December 2020; pp. 11.4.1–11.4.4.
5. Edelstein, D.; Rizzolo, M.; Sil, D.; Dutta, A.; DeBrosse, J.; Wordeman, M.; Arceo, A.; Chu, I.C.; Demarest, J.; Edwards, E.R.J.; et al. A 14 nm Embedded STT-MRAM CMOS Technology. In Proceedings of the 2020 IEEE Electron Devices Meeting (IEDM), San Francisco, CA, USA, 12–18 December 2020; pp. 11.5.1–11.5.4.
6. Lee, T.Y.; Yamane, K.; Otani, Y.; Zeng, D.; Kwon, J.; Lim, J.H.; Naik, V.B.; Hau, L.Y.; Chao, R.; Chung, N.L.; et al. Advanced MTJ Stack Engineering of STT-MRAM to Realize High Speed Applications. In Proceedings of the 2020 IEEE Electron Devices Meeting (IEDM), San Francisco, CA, USA, 12–18 December 2020; pp. 11.6.1–11.6.4.
7. Apalkov, D.; Dieny, B.; Slaughter, J.M. Magnetoresistive Random Access Memory. *Proc. IEEE* **2016**, *104*, 1796–1830. [CrossRef]
8. Hu, G.; Nowak, J.J.; Gottwald, M.G.; Brown, S.L.; Doris, B.; D'Emic, C.P.; Hashemi, P.; Houssameddine, D.; He, Q.; Kim, D.; et al. Spin-Transfer Torque MRAM with Reliable 2 ns Writing for Last Level Cache Applications. In Proceedings of the 2019 IEEE Electron Devices Meeting (IEDM), San Francisco, CA, USA, 7–11 December 2019; pp. 2.6.1–2.6.4.
9. Alzate, J.G.; Arslan, U.; Bai, P.; Brockman, J.; Chen, Y.J.; Das, N.; Fischer, K.; Ghani, T.; Heil, P.; Hentges, P.; et al. 2 MB Array-Level Demonstration of STT-MRAM Process and Performance Towards L4 Cache Applications. In Proceedings of the 2019 IEEE Electron Devices Meeting (IEDM), San Francisco, CA, USA, 7–11 December 2019; pp. 2.4.1–2.4.4.
10. Sakhare, S.; Perumkunnil, M.; Bao, T.H.; Rao, S.; Kim, W.; Crotti, D.; Yasin, F.; Couet, S.; Swerts, J.; Kundu, S.; et al. Enablement of STT-MRAM as Last Level Cache for the High Performance Computing Domain at the 5nm Node. In Proceedings of the 2018 IEEE Electron Devices Meeting (IEDM), San Francisco, CA, USA, 1–5 December 2018; pp. 18.3.1–18.3.4.
11. Aggarwal, S.; Almasi, H.; DeHerrera, M.; Hughes, B.; Ikegawa, S.; Janesky, J.; Lee, H.K.; Lu, H.; Mancoff, B.; Nagel, K.; et al. Demonstration of a Reliable 1Gb Standalone Spin-Transfer Torque MRAM for Industrial Applications. In Proceedings of the 2019 IEEE Electron Devices Meeting (IEDM), San Francisco, CA, USA, 7–11 December 2019; pp. 2.1.1–2.1.4.
12. Sato, H.; Honjo, H.; Watanabe, T.; Niwa, M.; Koike, H.; Miura, S.; Saito, T.; Inoue, H.; Nasuno, T.; Tanigawa, T.; et al. 14ns Write Speed 128Mb Density Embedded STT-MRAM with Endurance > 10^{10} and 10yrs Retention 85 °C Using Novel Low Damage MTJ Integration Process. In Proceedings of the 2018 IEEE Electron Devices Meeting (IEDM), San Francisco, CA, USA, 1–5 December 2018; pp. 27.2.1–27.2.4.

13. Golonzka, O.; Alzate, J.G.; Arslan, U.; Bohr, M.; Bai, P.; Brockman, J.; Buford, B.; Connor, C.; Das, N.; Doyle, B.; et al. MRAM as Embedded Non-Volatile Memory Solution for 22FFL FinFet Technology. In Proceedings of the 2018 IEEE Electron Devices Meeting (IEDM), San Francisco, CA, USA, 1–5 December 2018; pp. 18.1.1–18.1.4.
14. Miron, I.M.; Gaudin, G.; Auffret, S.; Rodmacq, B.; Schuhl, A.; Pizzini, S.; Vogel, J.; Gambardella, P. Current-Driven Spin Torque Induced by the Rashba Effect in a Ferromagnetic Metal Layer. *Nat. Mater.* **2010**, *9*, 230–234. [CrossRef] [PubMed]
15. Honjo, H.; Nguyen, T.V.A.; Watanabe, T.; Nasuno, T.; Zhang, C.; Tanigawa, T.; Miura, S.; Inoue, H.; Niwa, M.; Yoshiduka, T.; et al. First Demonstration of Field-Free SOT-MRAM with 0.35ns Write Speed and 70 Thermal Stability under 400 °C Thermal Tolerance by Canted SOT Structure and its Advanced Patterning/SOT Channel Technology. In Proceedings of the 2019 IEEE Electron Devices Meeting (IEDM), San Francisco, CA, USA, 7–11 December 2019; pp. 28.5.1–28.5.4.
16. Garello, K.; Yasin, F.; Hody, H.; Couet, S.; Souriau, L.; Sharifi, S.H.; Swerts, J.; Carpenter, R.; Rao, S.; Kim, W.; et al. Manufacturable 300 mm Platform Solution for Field-Free Switching SOT-MRAM. In Proceedings of the 2019 IEEE Symposium on VLSI Circuits, Kyoto, Japan, 9–14 June 2019; pp. T194–T195.
17. Garello, K.; Yasin, F.; Couet, S.; Souriau, L.; Swerts, J.; Rao, S.; Van Beek, S.; Kim, W.; Liu, E.; Kundu, S.; et al. SOT-MRAM 300 mm Integration for Low Power and Ultrafast Embedded Memories. In Proceedings of the 2018 IEEE Symposium on VLSI Circuits, Honolulu, HI, USA, 18–22 June 2018; pp. 81–82.
18. Fukami, S.; Anekawa, T.; Zhang, C.; Ohno, H. A Spin-Orbit Torque Switching Scheme with Collinear Magnetic Easy Axis and Current Configuration. *Nat. Nanotechnol.* **2016**, *11*, 621–626. [CrossRef] [PubMed]
19. Fukami, S.; Zhang, C.; DuttaGupta, S.; Kurenkov, A.; Ohno, H. Magnetization Switching by Spin-Orbit Torque in an Antiferromagnet-Ferromagnet Bilayer System. *Nat. Mater.* **2016**, *15*, 535–541. [CrossRef] [PubMed]
20. Oh, Y.W.; Baek, S.H.C.; Kim, Y.M.; Lee, H.Y.; Lee, K.D.; Yang, C.G.; Park, E.S.; Lee, K.S.; Kim, K.W.; Go, G.; et al. Field-Free Switching of Perpendicular Magnetization through Spin-Orbit Torque in Antiferromagnet/Ferromagnet/Oxide Structures. *Nat. Nanotechnol.* **2016**, *11*, 878–884. [CrossRef] [PubMed]
21. Wu, H.; Razavi, S.A.; Shao, Q.; Li, X.; Wong, K.L.; Liu, Y.; Yin, G.; Wang, K.L. Spin-Orbit Torque from a Ferromagnetic Metal. *Phys. Rev. B* **2019**, *99*, 184403. [CrossRef]
22. MacNeill, D.; Stiehl, G.M.; Guimaraes, M.H.D.; Buhrman, R.A.; Park, J.; Ralph, D.C. Control of Spin-Orbit Torques through Crystal Symmetry in WTe2/Ferromagnet Bilayers. *Nat. Phys.* **2016**, *13*, 300–305. [CrossRef]
23. Yu, G.; Upadhyaya, P.; Fan, Y.; Alzate, J.G.; Jiang, W.; Wong, K.L.; Takei, S.; Bender, S.A.; Chang, L.T.; Jiang, Y.; et al. Switching of Perpendicular Magnetization by Spin-Orbit Torques in the Absence of External Magnetic Fields. *Nat. Nanotechnol.* **2014**, *9*, 548–554. [CrossRef] [PubMed]
24. Sverdlov, V.; Makarov, A.; Selberherr, S. Two-Pulse Sub-ns Switching Scheme for Advanced Spin-Orbit Torque MRAM. *Solid-State Electron.* **2019**, *155*, 49–56. [CrossRef]
25. de Orio, R.L.; Ender, J.; Fiorentini, S.; Goes, W.; Selberherr, S.; Sverdlov, V. Numerical Analysis of Deterministic Switching of a Perpendicularly Magnetized Spin-Orbit Torque Memory Cell. *IEEE J. Electron Devices Soc.* **2021**, *9*, 61–67. [CrossRef]
26. Mehta, P.; Bukov, M.; Wang, C.H.; Day, A.G.; Richardson, C.; Fisher, C.K.; Schwab, D.J. A High-Bias, Low-Variance Introduction to Machine Learning for Physicists. *Phys. Rep.* **2019**, *810*, 1–124. [CrossRef] [PubMed]
27. Kovacs, A.; Fischbacher, J.; Oezelt, H.; Gusenbauer, M.; Exl, L.; Bruckner, F.; Suess, D.; Schrefl, T. Learning Magnetization Dynamics. *J. Magn. Magn. Mater.* **2019**, *491*, 165588. [CrossRef]
28. Exl, L.; Fischbacher, J.; Kovacs, A.; Oezelt, H.; Gusenbauer, M.; Yokota, K.; Shoji, T.; Hrkac, G.; Schrefl, T. Magnetic Microstructure Machine Learning Analysis. *J. Phys. Mater.* **2019**, *2*, 014001. [CrossRef]
29. Dai, M.; Hu, J.-M. Field-Free Spin-Orbit Torque Perpendicular Magnetization Switching in Ultrathin Nanostructures. *NPJ Comput. Mater.* **2020**, *6*, 78. [CrossRef]
30. Sutton, R.S.; Barto, A.G. *Reinforcement Learning: An Introduction*, 2nd. ed.; The MIT Press: Cambridge, MA, USA, 2018.
31. Mnih, V.; Silver, D.; Rusu, A.A.; Veness, J.; Bellemare, M.G.; Graves, A.; Riedmiller, M.; Fidjeland, A.K.; Ostrovski, G.; Petersen, S.; et al. Human-Level Control Through Deep Reinforcement Learning. *Nature* **2015**, *518*, 529–533. [CrossRef] [PubMed]
32. Silver, D.; Hubert, T.; Schrittwieser, J.; Antonoglou, I.; Lai, M.; Guez, A.; Lanctot, M.; Sifre, L.; Kumaran, D.; Graepel, T.; et al. A General Reinforcement Learning Algorithm that Masters Chess, Shogi, and Go Through Self-Play. *Science* **2018**, *362*, 1140–1144. [CrossRef] [PubMed]
33. de Orio, R.L.; Makarov, A.; Goes, W.; Ender, J.; Fiorentini, S.; Sverdlov, V. Two-Pulse Magnetic Field-Free Switching Scheme for Perpendicular SOT-MRAM with a Symmetric Square Free Layer. *Phys. B Condens. Matter* **2020**, *578*, 411743. [CrossRef]
34. Makarov, A. Modeling of Emerging Resistive Switching Based Memory Cells. Ph.D. Thesis, Institute for Microelectronics, TU Wien, Austria, 2014. Available online: https://www.iue.tuwien.ac.at/phd/makarov/ (accessed on 12 March 2021).
35. Raffin, A.; Hill, A.; Ernestus, M.; Gleave, A.; Kanervisto, A.; Dormann, N. Stable Baselines 3. Available online: https://github.com/DLR-RM/stable-baselines3 (accessed on 12 March 2021).

micromachines

MDPI

Article

Theoretical Study of Field-Free Switching in PMA-MTJ Using Combined Injection of STT and SOT Currents

Shaik Wasef and Hossein Fariborzi *

Electrical and Computer Engineering, King Abdullah University of Science and Technology, Thuwal 23955, Saudi Arabia; shaik.wasef@kaust.edu.sa
* Correspondence: hossein.fariborzi@kaust.edu.sa

Abstract: Field-free switching in perpendicular magnetic tunnel junctions (P-MTJs) can be achieved by combined injection of spin-transfer torque (STT) and spin-orbit torque (SOT) currents. In this paper, we derived the relationship between the STT and SOT critical current densities under combined injection. We included the damping–like torque (DLT) and field-like torque (FLT) components of both the STT and SOT. The results were derived when the ratio of the FLT to the DLT component of the SOT was positive. We observed that the relationship between the critical SOT and STT current densities depended on the damping constant and the magnitude of the FLT component of the STT and the SOT current. We also noted that, unlike the FLT component of SOT, the magnitude and sign of the FLT component of STT did not have a significant effect on the STT and SOT current densities required for switching. The derived results agreed well with micromagnetic simulations. The results of this work can serve as a guideline to model and develop spintronic devices using a combined injection of STT and SOT currents.

Keywords: combined spin-transfer torque (STT) and spin-orbit torque (SOT) switching; field like torque; damping like torque; magnetic tunnel junction

Citation: Wasef, S.; Fariborzi, H. Theoretical Study of Field-Free Switching in PMA-MTJ Using Combined Injection of STT and SOT Currents. *Micromachines* **2021**, *12*, 1345. https://doi.org/10.3390/mi12111345

Academic Editors: Viktor Sverdlov and Nuttachai Jutong

Received: 28 August 2021
Accepted: 26 October 2021
Published: 31 October 2021

1. Introduction

Information can be stored in ferromagnetic structures by the interaction between spin-polarized currents and magnetic moments. An magnetic tunnel junctions (MTJ) consists of a tunneling oxide layer (usually MgO) deposited between two ferromagnetic layers. Binary information is stored based on the relative orientation of the free layer (FL) to the reference layer (RL). An antiparallel (AP) orientation offers a high resistance and a parallel (P) orientation offers low resistance. Usually, the AP state is used to store bit "1" and the P state is used to store bit "0". The AP or P state can be obtained by the interaction of the FL with spin-polarized charges. Depending on the mechanism of interaction, the magnetic storage devices can be classified into spin-transfer torque (STT) devices and spin-orbit torque (SOT) devices. In STT devices (Figure 1a), spin-polarized charges are generated via spin filtering from the RL of the MTJ. These charges can transfer their spin angular momentum to the FL, thereby exerting torque on its magnetization, which changes its magnetic orientation [1–3]. In SOT (Figure 1b), the magnetization switching in the free layer takes place due to the surface (Rashba effect) and bulk interactions (spin hall effect) caused by the attached heavy metal layer [4–6]. The magnetic reversal in the aforementioned mechanisms is due to the combined effects of DLT and FLT vector components [7–10]. In fact, the FLT component can affect the critical current required for switching in both STT and SOT devices [11,12]. Although commonly used, STT devices suffer from reliability and endurance issues caused by damage to the thin MgO tunneling layer. This happens because of the repeated tunneling of electrons, as the read and write paths are overlapped (both out of plane) [13,14]. In addition to this, an STT device suffers from incubation delay and, unlike SOT, does not realize sub-nanosecond switching [15]. On the other hand, an SOT device requires an external in-plane bias field for deterministic switching [16]. In order to overcome these constraints, devices operating under the combined effects of STT and SOT have been experimentally demonstrated [17]. The use of combined injection of STT and SOT currents provides a two-way advantage. The use of an STT current component facilitates complete magnetic reversal, which would otherwise require an external bias field in an SOT device. On the other hand, the SOT current component can provide lower switching time than a pure STT device. Due to these advantages, it was deemed necessary to comprehensively analyze the behavior of STT-SOT devices (Figure 1c). Although these devices have been extensively studied through macrospin simulations [18–20], their analysis under the influence of DLT and FLT has yet to be explored.

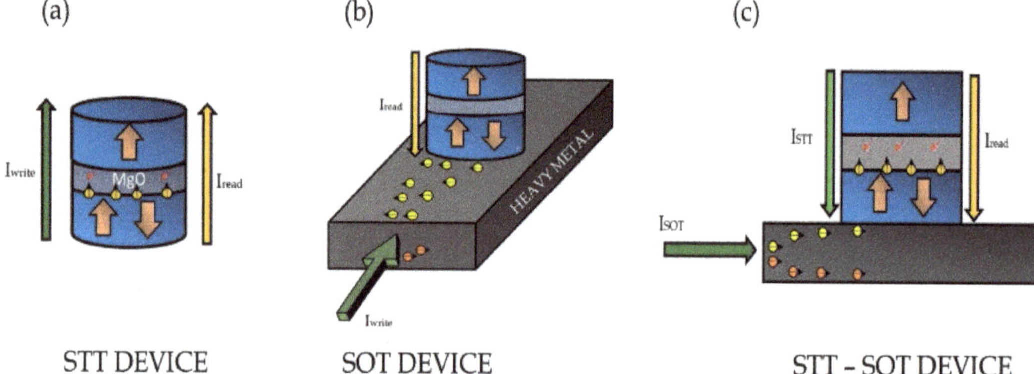

Figure 1. A schematic of the (**a**) spin-transfer torque (STT) device (**b**) spin-orbit torque (SOT) device and (**c**) STT-SOT device.

In this paper, we investigated the effects of combined injection of SOT (J_{SOT}) and STT (J_{STT}) current in P-MTJs with their individual DLT and FLT components under zero bias field. We first derived the critical STT density ($J^{STT}_{critical}$), required for switching in the absence of any SOT current. We then derived the relationship between the STT and SOT critical current densities when the ratio of the FLT to the DLT component of the SOT (β_{SOT}) was positive. We observed that, under combined injection, the critical SOT current density depended on damping constant and the magnitude of the FLT component of the STT current and the SOT current. We also noted that the critical STT and SOT current densities required for switching did not change considerably with the magnitude and sign of the FLT component of STT. However, they decreased with the increasing magnitude of FLT component of SOT. The derived results were verified with a micromagnetic model developed in OOMMF [21].

2. Landau–Lifshitz–Gilbert Equation with Spin-Transfer Torque (STT) and Spin-Orbit Torque (SOT) Terms

The magnetization dynamics of a ferromagnet under the influence of magnetic fields (internal and external) and spin currents can be described by the LLG equation with additional STT and SOT terms as given below [3].

$$\frac{d\vec{m}}{dt} = -\gamma\left(\vec{m} \times \vec{H}\right) + \alpha\left(\vec{m} \times \frac{d\vec{m}}{dt}\right) + \vec{\tau}_{DL-SOT} + \vec{\tau}_{FL-SOT} + \vec{\tau}_{DL-STT} + \vec{\tau}_{FL-STT} \quad (1)$$

$$\vec{\tau}_{DL-SOT} = -\gamma H_{SOT}\left(\vec{m} \times \left(\hat{p}_{SOT} \times \vec{m}\right)\right)$$

$$\vec{\tau}_{FL-SOT} = -\gamma \beta_{SOT} H_{SOT}\left(\vec{m} \times \hat{p}_{SOT}\right)$$

$$\vec{\tau}_{DL-STT} = -\gamma H_{STT}\left(\vec{m} \times \left(\hat{p}_{STT} \times \vec{m}\right)\right)$$

$$\vec{\tau}_{FL-STT} = -\gamma\beta_{STT}H_{STT}\left(\vec{m} \times \hat{p}_{STT}\right)$$

Here, γ is the gyromagnetic ratio, β_{STT} (β_{SOT}) is the ratio of the FLT to DLT of the STT (SOT), α is the damping constant, $\vec{m}$ is the unit vector which represents the magnetic orientation of the FL, $\hat{p}_{STT}$ and $\hat{p}_{SOT}$ are the spin polarization directions, and H_{STT} and H_{SOT} are the spin torque strengths of the STT and SOT, respectively, described as follows:

$$H_{STT} = \frac{\hbar\eta J_{STT}}{2eM_s t_{FM}}$$

$$H_{SOT} = \frac{\hbar\theta_{SHE} J_{SOT}}{2eM_s t_{FM}}$$

Here, e is the electron charge, $\hbar$ is the reduced Planck's constant, η is the spin polarization constant, M_s is the saturation magnetization of the FL, θ_{SHE} is the spin hall angle, t_{FM} is the thickness of the free layer, and J_{STT} and J_{SOT} are the STT and SOT charge current densities, respectively.

For simplicity, we ignored the effect of the stray fields of the RL on the FL. We also ignored the effects of the Oersted fields generated by the STT and SOT currents, as they only provided an initial misalignment in the FL magnetization and did not contribute significantly toward switching [22]. The analysis and the micromagnetic simulations (refer to methods: micromagnetic model) were developed based on Equation (1).

Unless otherwise specified, parametric values adopted in this work are mentioned in Table 1.

Table 1. Input parameters used in this work unless otherwise specified.

Parameters	Numerical Values
γ	$17.32 \times 10^{11} radT^{-1}s^{-1}$
α	0.005
η	0.33
M_s	$1.5 \times 10^6 A/m$ [23]
t_{FM}	1 nm [23]
H_{Keff}	$540Oe$ [23]
$\theta_{SHE}(\beta - Ta)$	0.1^4
$\hat{p}_{STT}$	$\hat{e}_z$
$\hat{p}_{SOT}$	$\hat{e}_y$
β_{SOT}	2
β_{STT}	1
$A_{exchange}$	20 pJ/m
T_{rise} (J_{STT}, J_{SOT})	0.5 ns
T_{fall} (J_{STT}, J_{SOT})	0.5 ns

3. Results

COMBINED STT-SOT Induced Switching in PMA-MTJ

In this section, we theoretically derived the relationship between the STT and SOT current densities under combined injection. The relationship was derived for FL switching from P to AP state. However, the same approach could be extended to obtain the relationship for switching from AP to P. The duration of the STT pulse in simulations was kept larger than the SOT pulse to promote deterministic switching [17]. As evident from Equation (1), the magnetic destabilization in these devices took place under the influence of an effective field (refer Figure 2b) given by

$$\vec{H}_{eff} = \vec{H} + \beta_{SOT}H_{SOT}\hat{p}_{SOT} + \beta_{STT}H_{STT}\hat{p}_{STT} \quad (2)$$

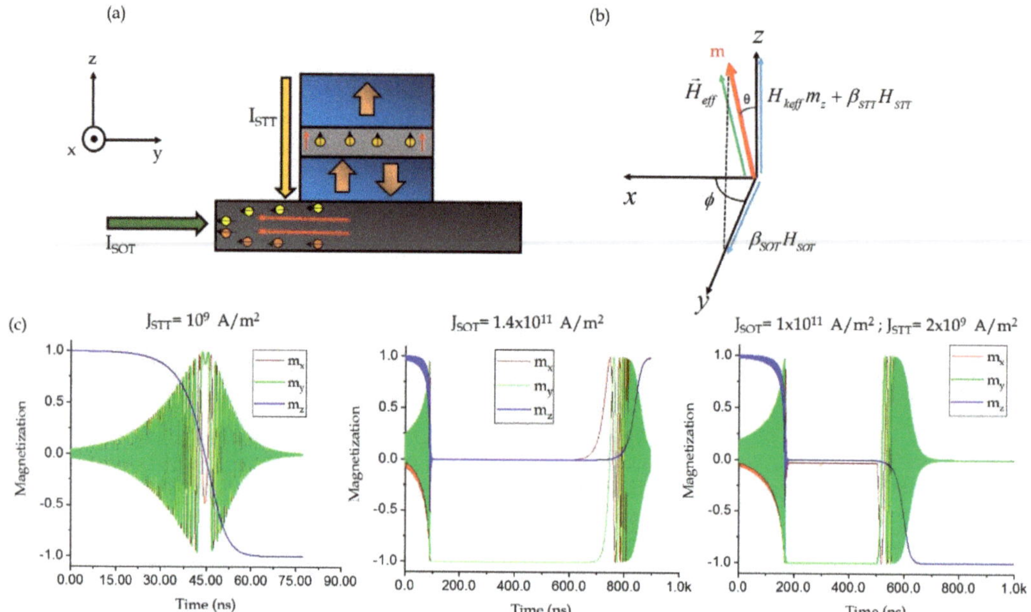

Figure 2. (**a**) Schematic of the STT-SOT device configuration used (**b**) Magnetization m relaxing to a point of equilibrium along the $\vec{H}_{eff}$ direction before reversal. (**c**) Magnetization dynamics of the FL in an STT, SOT and STT-SOT device.

Switching when $\beta_{SOT} > 0$ took place through precessions, since both STT and SOT directly compete with damping[12] (refer Figure 2c). Thus, we were able to derive the relation between J_{SOT} and J_{STT} by linearizing the LLG equation. The magnetization dynamics of the FL under combined injection, as described by Equation (1), can be modified to the following form:

$$-\left(\frac{1+\alpha^2}{\gamma}\right)\frac{d\vec{m}}{dt} = \left(\vec{m}\times\vec{H}\right) + \alpha\left(\vec{m}\times\left(\vec{m}\times\vec{H}\right)\right) - H_{STT}(\alpha\beta_{STT}-1)\left(\vec{m}\times\left(\hat{p}_{STT}\times\vec{m}\right)\right) + H_{STT}(\alpha+\beta_{STT})\left(\vec{m}\times\hat{p}_{STT}\right) - H_{SOT}(\alpha\beta_{SOT}-1)\left(\vec{m}\times\left(\hat{p}_{SOT}\times\vec{m}\right)\right) + H_{SOT}(\alpha+\beta_{SOT})\left(\vec{m}\times\hat{p}_{SOT}\right) \quad (3)$$

Equation (3) can be linearized by converting the coordinate's axes xyz to a new XYZ system where Z aligns with the direction of $\vec{H}_{eff}$ by using the rotation matrix R given by

$$R = \begin{pmatrix} \cos\theta\cos\phi & \cos\theta\sin\phi & -\sin\theta \\ -\sin\phi & \cos\phi & 0 \\ \sin\theta\cos\phi & \sin\theta\sin\phi & \cos\theta \end{pmatrix}$$

Here, θ and ϕ are the polar and azimuthal angles of the effective field when SOT and STT current approach their critical values (shown in Figure 2b). We linearized the LLG equation based on the assumption that the Z–component of magnetization remains unchanged at the beginning of the reversal and reversal occurs after small perturbations around the equilibrium direction. Thus, for simplification, we considered

$$\begin{cases} M_Z = 1 \\ M_Y.M_X << 1 \\ M_X^2, M_Y^2 = 0 \end{cases}$$

Using the above assumptions Equation (3) can be modified into the following form

$$\frac{1+\alpha^2}{\gamma}\begin{pmatrix} dM_X/dt \\ dM_Y/dt \end{pmatrix} = M\begin{pmatrix} M_X \\ M_Y \end{pmatrix} + G \tag{4}$$

Equation (4) has solutions of the form $M_X, M_Y = A\exp\left(-\gamma\left\{[\pm i\sqrt{|M| - (\text{Trace}[M]/2)^2} - \text{Trace}[M]/2]t\right\}\right)$, where the real part in the exponential represents the time evolution of the oscillation amplitude. Thus, the realization of switching was based on the boundary condition of Trace [M] = 0. Hence, we obtained

$$M_{11} + M_{22} = -2H_{keff}\alpha\cos^2\theta + H_{keff}\alpha\sin^2\theta + 2H_{SOT}(1-\alpha\beta_{SOT})\sin\phi\sin\theta + 2H_{STT}(1-\alpha\beta_{STT})cos\theta = 0 \tag{5}$$

Substituting the values of θ and ϕ (from supplementary note 1), we first derived the critical switching current density ($J^{STT}_{critical}$) for STT-based switching, as follows:

$$J^{STT}_{critical} = \frac{2et_{FM}M_s\alpha H_{keff}}{\hbar\eta(1-\alpha\beta_{STT})} \tag{6}$$

From Equation (6), we observed that $J^{STT}_{critical}$ depended on the magnitude and sign of β_{STT}. $J^{STT}_{critical}$ did not change significantly with increase in β_{STT}, as shown in Figure 3. This result was consistent with observations made by Carpentieri et al. [24]. In addition to this, the rate of increase was relatively $J^{STT}_{critical}$, with β_{STT} greater for larger values of α. The value of β_{STT} depended on the properties of the materials [7,25–30] and was experimentally estimated to be between 0.01–0.1 for a CoFeB/MgO/CoFeB [29,30]. In this article, we used β_{STT} values greater than the experimentally measured results to clearly show its effect. Here, a positive value of $J^{STT}_{critical}$ refers to the electrons moving from the FL to the RL.

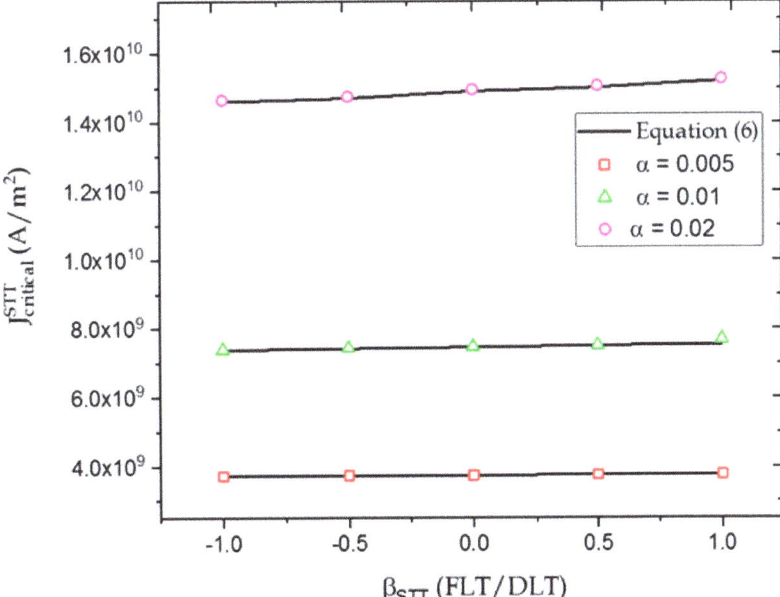

Figure 3. Dependence of $J^{STT}_{critical}$ on β_{STT} for α = 0.005, 0.01, 0.02. The solid lines and symbols represent the results obtained from Equation (6) and micromagnetic simulations respectively.

Including the effects of SOT in Equation (5), we determined the relationship between the critical STT and SOT current densities, above which the P-MTJ switched from P-AP state as follows

$$J_{SOT} = \frac{\sqrt{2}\sqrt{\alpha + \xi_{STT} J_{STT}(\alpha\beta_{STT} - 1)}(1 + \xi_{STT} J_{STT} \beta_{STT})}{\xi_{SOT}\sqrt{\beta_{SOT}(2 + \alpha\beta_{SOT} - \xi_{STT} J_{STT}(\beta_{SOT} - 2\beta_{STT} + \alpha\beta_{SOT}\beta_{STT}))}} \tag{7}$$

where $\xi_{STT} = \frac{\hbar\eta}{2et_{FM}M_s H_{keff}}$ and $\xi_{SOT} = \frac{\hbar\theta_{SHE}}{2et_{FM}M_s H_{keff}}$

Equation (7) is valid only when $\beta_{SOT} > 0$, since for $\beta_{SOT} = 0$, switching did not take place entirely through precessions, although the STT always competed with the damping torque (refer to supplementary note 2, (Figure S1)). In the absence of J_{STT}, Equation (7) was consistent with results obtained by Tanuguchi.et al. [12]. As seen in Figure 4a, the critical current densities did not decrease appreciably, even for very large values of β_{STT}. However, their magnitudes decreased considerably with increasing values of β_{SOT} (Figure 4b). This is because the FLT components of STT and SOT added to the effective field in the $\hat{e}_z$ direction and $\hat{e}_y$ direction, respectively (Equation (2)). Since the magnitude of J_{STT} required for switching was lower than J_{SOT}, the contribution of its FLT component to the effective field was insignificant. Additionally, the FLT component of STT did not contribute toward a significant tilt in the magnetization. On the contrary, the FLT component of SOT was stronger, owing to the large SOT current density. As the FLT component of SOT was in-plane, it provided a larger tilt to the magnetization from its initial position, thereby reducing the individual critical current for switching. Hence, J_{STT} and J_{SOT}, under combined injection, decreased appreciably for increasing values of β_{SOT}. Here, positive values of J_{STT} refer to electrons flowing from FL to RL and positive values of J_{SOT} refer to electrons flowing in the negative $\hat{e}_y$ direction (refer Figure 2a). It must be noted that deterministic switching took place only in the presence of combined STT and SOT and did not take place in the presence of SOT alone.

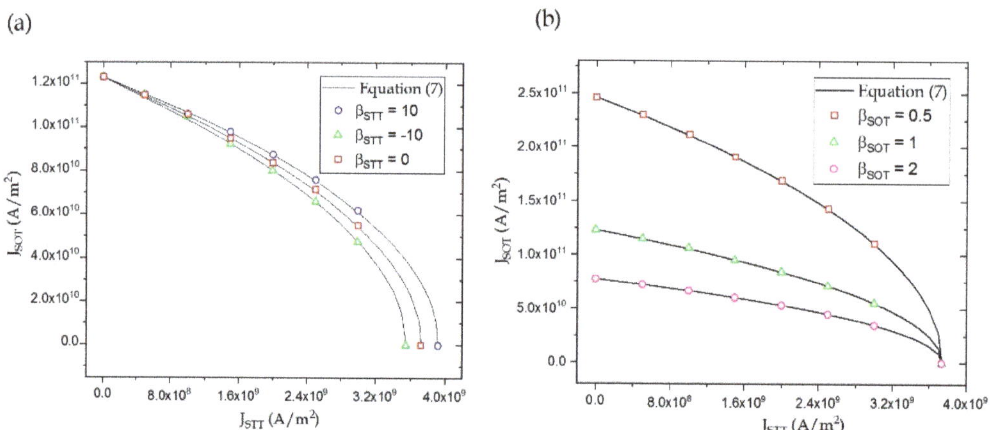

Figure 4. The solid line represents boundary Equation (7) above switching takes place from P to AP state (**a**), with changing β_{STT} and (**b**) with increasing β_{SOT}. Symbols represent results obtained from micromagnetic simulations.

SOT switching is symmetric in nature, since the final configuration of the FL is in-plane irrespective of the direction of current injection. Unlike SOT, STT-based switching is asymmetric, i.e., the magnitude of J_{STT} for AP to P switching is lower than J_{STT} required for P to AP switching. However, this inclusion was beyond the scope of this work. Figure 5 shows the boundaries separating the different regions of switching for parameters mentioned in Table 1. As seen in Figure 5, Equation (7) was consistent the experimental results obtained by Wang et al. [17].

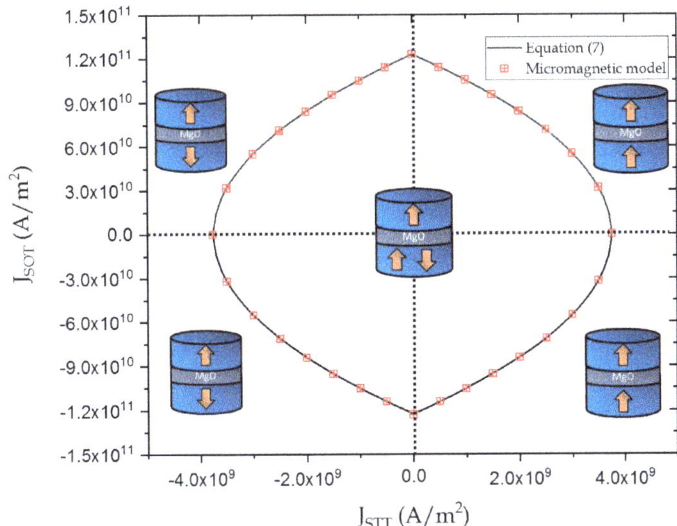

Figure 5. The solid line represents the boundary Equation (7). The symbols represent the results obtained from micromagnetic simulations.

4. Conclusions

In this work, we investigated the magnetic switching in MTJ devices under combined injection of Spin transfer torque (STT) and Spin orbit torque (SOT) currents. We included the effects of both the damping-like and field-like torque of the STT and SOT currents. We derived the relationship between the STT and SOT current densities when the ratio of the FLT to DLT component of the SOT was positive. We observed that the relationship between the critical SOT and STT current densities under combined injection depended on the damping constant and the magnitude of the FLT component of the STT current and the SOT current. However, unlike the FLT component of SOT, the magnitude and sign of the FLT component of STT had an insignificant effect on the STT and SOT current densities. The derived results were verified with a micromagnetic model.

5. Methods

Micromagnetic Model

In this work, the micro-magnetic model was developed in OOMMF [21] based on Equation (1). Combined injection of STT and SOT was implemented using the "Oxs_SpinXferEvolve" extension module. The field-like torque components of STT and SOT were added as external magnetic fields with magnitudes depending on the individual injection currents. The duration of the STT current pulse was kept larger than the SOT to promote deterministic switching [17].

Supplementary Materials: The following are available online at https://www.mdpi.com/article/10.3390/mi12111345/s1, Figure S1: Magnetic switching under combined injection of STT and SOT when $\beta_{SOT} = 0$.

Author Contributions: Formal analysis, S.W.; Project administration, S.W.; Writing—original draft, S.W.; Writing—review & editing, H.F. All authors contributed equally to the analysis and writing of this article. All authors have read and agreed to the published version of the manuscript.

Funding: This research received no external funding.

Conflicts of Interest: The authors declare no competing interests.

References

1. Hu, G.; Lee, J.H.; Nowak, J.J.; Sun, J.Z.; Harms, J.; Annunziata, A.; Brown, S.; Chen, W.; Kim, Y.H.; Lauer, G.; et al. STT-MRAM with double magnetic tunnel junctions. In Proceedings of the IEEE 2015 Internaional Electron Devices Meeting (IEDM), Washington, DC, USA, 7–9 December 2015.
2. Ikeda, S.; Sato, H.; Honjo, H.; Enobio, E.C.I.; Ishikawa, S.; Yamanouchi, M.; Fukami, S.; Kanai, S.; Matsukura, F.; Endoh, T.; et al. Perpendicular-anisotropy CoFeB-MgO based magnetic tunnel junctions scaling down to 1X nm. In Proceedings of the 2014 IEEE International Electron Devices Meeting, San Francisco, CA, USA, 15–17 December 2014.
3. Slonczewski, J.C. Current-driven excitation of magnetic multilayers. *J. Magn. Magn. Mater.* **1996**, *159*, L1–L7. [CrossRef]
4. Liu, L.; Pai, C.-F.; Li, Y.; Tseng, H.W.; Ralph, D.C.; Buhrman, R.A. Spin-torque switching with the giant spin Hall effect of tantalum. *Science* **2012**, *336*, 555–558. [CrossRef]
5. Miron, I.M.; Gaudin, G.; Auffret, S.; Rodmacq, B.; Schuhl, A.; Pizzini, S.; Vogel, J.; Gambardella, P. Current-driven spin torque induced by the Rashba effect in a ferromagnetic metal layer. *Nat. Mater.* **2010**, *9*, 230–234. [CrossRef]
6. Miron, I.M.; Garello, K.; Gaudin, G.; Zermatten, P.-J.; Costache, M.V.; Auffret, S.; Bandiera, S.; Rodmacq, B.; Schuhl, A.; Gambardella, P. Perpendicular switching of a single ferromagnetic layer induced by in-plane current injection. *Nature* **2011**, *476*, 189–193. [CrossRef] [PubMed]
7. Zhang, S.; Levy, P.M.; Fert, A. Mechanisms of Spin-Polarized Current-Driven Magnetization Switching. *Phys. Rev. Lett.* **2002**, *88*, 236601. [CrossRef] [PubMed]
8. Heinonen, O.G.; Stokes, S.W.; Yi, J.Y. Perpendicular spin torque in magnetic tunnel junctions. *Phys. Rev. Lett.* **2010**, *105*, 066602. [CrossRef] [PubMed]
9. Kim, J.; Sinha, J.; Hayashi, M.; Yamanouchi, M.; Fukami, S.; Suzuki, T.; Mitani, S.; Ohno, H. Layer thickness dependence of the current-induced effective field vector in Ta/CoFeB/MgO. *Nat. Mater.* **2012**, *12*, 240–245. [CrossRef] [PubMed]
10. Zhang, C.; Yamanouchi, M.; Sato, H.; Fukami, S.; Ikeda, S.; Matsukura, F.; Ohno, H. Magnetotransport measurements of current induced effective fields in Ta/CoFeB/MgO. *Appl. Phys. Lett.* **2013**, *103*, 262407. [CrossRef]
11. Zhou, Y. Effect of the field-like spin torque on the switching current and switching speed of magnetic tunnel junction with perpendicularly magnetized free layers. *J. Appl. Phys.* **2011**, *109*, 023916. [CrossRef]
12. Taniguchi, T.; Mitani, S.; Hayashi, M. Critical current destabilizing perpendicular magnetization by the spin Hall effect. *Phys. Rev. B* **2015**, *92*, 024428. [CrossRef]
13. Chun, K.C.; Zhao, H.; Harms, J.D.; Kim, T.H.; Wang, J.-P.; Kim, C.H. A scaling roadmap and performance evaluation of in-plane and perpendicular MTJ based STT-MRAMs for high-density cache memory. *IEEE J. Solid-State Circuits* **2013**, *48*, 598–610. [CrossRef]
14. Zhao, W.; Zhang, Y.; Devolder, T.; Klein, J.-O.; Ravelosona, D.; Chappert, C.; Mazoyer, P. Failure and reliability analysis of STT-MRAM. *Microelectron. Reliab.* **2012**, *52*, 1848–1852. [CrossRef]
15. Garello, K.; Avci, C.O.; Miron, I.M.; Baumgartner, M.; Ghosh, A.; Auffret, S.; Boulle, O.; Gaudin, G.; Gambardella, P. Ultrafast magnetization switching by spin-orbit torques. *Appl. Phys. Lett.* **2014**, *105*, 212402. [CrossRef]
16. Garello, K.; Avci, C.O.; Miron, I.M.; Baumgartner, M.; Ghosh, A.; Auffret, S.; Boulle, O.; Gaudin, G.; Gambardella, P. Spin–orbit torque magnetization switching of a three-terminal perpendicular magnetic tunnel junction. *Appl. Phys. Lett.* **2014**, *104*, 042406.
17. Wang, M.; Cai, W.; Zhu, D.; Wang, Z.; Kan, J.; Zhao, Z.; Cao, K.; Wang, Z.; Zhang, Y.; Zhang, T.; et al. Field-free switching of a perpendicular magnetic tunnel junction through the interplay of spin–orbit and spin-transfer torques. *Nat. Electron.* **2018**, *1*, 582–588. [CrossRef]
18. Wang, Z.; Zhao, W.; Deng, E.; Zhang, Y.; Klein, J.O. Magnetic non-volatile flip-flop with spin-Hall assistance. *Phys. Status Solidi-Rapid Res. Lett.* **2015**, *9*, 375–378. [CrossRef]
19. Wang, Z.; Zhao, W.; Deng, E.; Klein, J.; Chappert, C. Perpendicular anisotropy magnetic tunnel junction switched by spin-Hall-assisted spin-transfer torque. *J. Phys. D* **2015**, *48*, 065001. [CrossRef]
20. Brink, A.A.V.D.; Cosemans, S.; Cornelissen, S.; Manfrini, M.; Vaysset, A.; Van Roy, W.; Min, T.; Swagten, H.J.M.; Koopmans, B.B. Spin-Hall-assisted magnetic random access memory. *Appl. Phys. Lett.* **2014**, *104*, 012403. [CrossRef]
21. Donahue, M.J.; Porter, D.G. (Eds.) *OOMMF User's Guide, Version 1.0*; Interagency Report NISTIR 6376; National Institute of Standards and Technology: Gaithersburg, MD, USA, 1999.
22. You, C.-Y. Micromagnetic Simulations for Spin Transfer Torque in Magnetic Multilayers. *J. Magn.* **2012**, *17*, 73–77. [CrossRef]
23. Liu, L.; Lee, O.J.; Gudmundsen, T.J.; Ralph, D.C.; Buhrman, R.A. Current-Induced switching of perpendicularly magnetized magnetic layers using spin torque from the spin Hall effect. *Phys. Rev. Lett.* **2012**, *109*, 096602. [CrossRef]
24. Siracusano, G.; Tomasello, R.; d'Aquino, M.; Puliafito, V.; Giordano, A.; Azzerboni, B.; Braganca, P.; Finocchio, G.; Carpentieri, M. Description of statistical switching in perpendicular STT-MRAM within an analytical and numerical micromagnetic framework. *IEEE Trans. Magn.* **2018**, *54*, 1400210. [CrossRef]
25. Zwierzycki, M.; Tserkovnyak, Y.; Kelly, P.J.; Brataas, A.; Bauer, G.E.W. First-principles study of magnetization relaxation enhancement and spin transfer in thin magnetic films. *Phys. Rev. B* **2005**, *71*, 064420. [CrossRef]
26. Theodonis; Kioussis, N.; Kalitsov, A.; Chshiev, M.; Butler, W.H. Anomalous bias dependence of spin torque in magnetic tunnel junctions. *Phys. Rev. Lett.* **2006**, *97*, 237205. [CrossRef] [PubMed]
27. Xiao, J.; Bauer, G.E.W. Spin-transfer torque in magnetic tunnel junctions: Scattering theory. *Phys. Rev. B* **2008**, *77*, 224419. [CrossRef]

28. Tulapurkar, A.A.; Suzuki, Y.; Fukushima, A.; Kubota, H.; Maehara, H.; Tsunekawa, K.; Djayaprawira, D.D.; Watanabe, N.; Yuasa, S. Spin-torque diode effect in magnetic tunnel junctions. *Nature* **2005**, *438*, 339. [CrossRef]
29. Kubota, H.; Fukushima, A.; Yakushiji, K.; Nagahama, T.; Yuasa, S.; Ando, K.; Maehara, H.; Nagamine, Y.; Tsunekawa, K.; Djayaprawira, D.D.; et al. Quantitative measurement of voltage dependence of spin-transfer torque in MgO-based magnetic tunnel junctions. *Nat. Phys.* **2008**, *4*, 37. [CrossRef]
30. Sankey, J.C.; Cui, Y.-T.; Sun, J.Z.; Slonczewski, J.C.; Buhrman, R.A.; Ralph, D.C. Measurement of the spin-transfer-torque vector in magnetic tunnel junctions. *Nat. Phys.* **2008**, *4*, 67. [CrossRef]

MDPI
St. Alban-Anlage 66
4052 Basel
Switzerland
Tel. +41 61 683 77 34
Fax +41 61 302 89 18
www.mdpi.com

Micromachines Editorial Office
E-mail: micromachines@mdpi.com
www.mdpi.com/journal/micromachines

www.ingramcontent.com/pod-product-compliance
Lightning Source LLC
LaVergne TN
LVHW070610170726
843515LV00004B/35

* 9 7 8 3 0 3 6 5 3 8 4 2 6 *